**Springer Theses**

Recognizing Outstanding Ph.D. Research

More information about this series at http://www.springer.com/series/8790

## Aims and Scope

The series "Springer Theses" brings together a selection of the very best Ph.D. theses from around the world and across the physical sciences. Nominated and endorsed by two recognized specialists, each published volume has been selected for its scientific excellence and the high impact of its contents for the pertinent field of research. For greater accessibility to non-specialists, the published versions include an extended introduction, as well as a foreword by the student's supervisor explaining the special relevance of the work for the field. As a whole, the series will provide a valuable resource both for newcomers to the research fields described, and for other scientists seeking detailed background information on special questions. Finally, it provides an accredited documentation of the valuable contributions made by today's younger generation of scientists.

## Theses are accepted into the series by invited nomination only and must fulfill all of the following criteria

- They must be written in good English.
- The topic should fall within the confines of Chemistry, Physics, Earth Sciences, Engineering and related interdisciplinary fields such as Materials, Nanoscience, Chemical Engineering, Complex Systems and Biophysics.
- The work reported in the thesis must represent a significant scientific advance.
- If the thesis includes previously published material, permission to reproduce this must be gained from the respective copyright holder.
- They must have been examined and passed during the 12 months prior to nomination.
- Each thesis should include a foreword by the supervisor outlining the significance of its content.
- The theses should have a clearly defined structure including an introduction accessible to scientists not expert in that particular field.

Iman Tavassoly

# Dynamics of Cell Fate Decision Mediated by the Interplay of Autophagy and Apoptosis in Cancer Cells

## Mathematical Modeling and Experimental Observations

Doctoral thesis accepted by Virginia Polytechnic Institute and State University, USA

Springer

Iman Tavassoly
Program of Genetics, Bioinformatics
and Computational Biology (GBCB)
Virginia Polytechnic Institute
and State University
Blacksburg, VA, USA

Additional material to this book can be downloaded from http://extras.springer.com.

ISSN 2190-5053 ISSN 2190-5061 (electronic)
Springer Theses
ISBN 978-3-319-14961-5 ISBN 978-3-319-14962-2 (eBook)
DOI 10.1007/978-3-319-14962-2

Library of Congress Control Number: 2015930194

Springer Cham Heidelberg New York Dordrecht London


Printed on acid-free paper

Springer International Publishing AG Switzerland is part of Springer Science+Business Media (www.springer.com)

A part of this thesis is under revision in CPT: Pharmacometrics and Systems Pharmacology.

This work received support from US National Institutes of Health grants U54-CA149147 and Transdisciplinary Team Science (TTS) Fellowship from Virginia Bioinformatics Institute (VBI).

*This dissertation is dedicated to Ludwig Wittgenstein because my mind was nourished by his philosophy and ideas during 4 years of my Ph.D. at Virginia Tech. He had written in a letter to a friend : "....I wish you could live quiet, in a sense, and be in a position to be kind and understanding to all sorts of human beings who need it. Because we all need this sort of thing very badly". (Malcolm, Norman. "Ludwig Wittgenstein: a memoir." (2001).)*

*This dissertation is also dedicated to Elnaz Farbod who has lived based on what Wittgenstein has mentioned above.*

# Foreword

Iman Tavassoly's Ph.D. thesis presents the first thorough theoretical study of the interplay between autophagy and apoptosis in the response of cancer cells to starvation or chemotoxic stress. In Chap. 1, Dr. Tavassoly reviews the experimental facts about cell death modalities, autophagy and apoptosis, and the molecular mechanisms of these pathways and their interactions. In Chap. 2, he builds a mathematical model of the molecular mechanisms and compares simulations of the model to experimental observations of how rat kidney cells respond to cisplatin treatment. In Chap. 3, he presents single-cell measurements of the dynamics of the autophagic response of human neuroglioma cells to cisplatin treatment and of human breast cancer cell lines to serum starvation. Dr. Tavassoly's thesis is a fine example of applying theoretical and experimental methods of molecular systems biology to gain a deeper understanding of an important problem in human health.

Blacksburg, VA

John J. Tyson
University Distinguished Professor
of Molecular Systems Biology

# Acknowledgments

When I was a medical student, being a computational and systems biologist was a dream for me and this dream became a reality after joining Dr. John Tyson's Lab at Virginia Tech. John really had a great impact on my transition from being a physician to be a mathematical and systems biologist. What I learned from John was not just about systems biology. I learned how to approach a question in science from him. Being in John's lab has changed the way I look at scientific questions in biology and medicine and I will always use the approaches I learned from him in my future works. John is a visionary leader in systems biology and I believe the methodologies he has developed will be widely used as common methodologies in biology. I would like to thank John very much for being my mentor and for all of his support.

I would also like to thank all the members of Tyson Lab, especially Jignesh Parmar, Anael Verdugo, Chun Chen, Dr. Yan Fu, Dr. Sandip Kar, Dr. Rajat Singhania, Dr. Baris Hancioglu, Kartik Subramanian, Dr. Teeraphan Laomettachit, and Dr. Janani Ravi.

I would like to thank Dr. Robert Clarke for his mentoring and kind support regarding my project on live-cell imaging of autophagy. I spent 1 year in his lab at Georgetown Lombardi Comprehensive Cancer Center and that 1 year was a turning point in my scientific career as I started to work on experimental methods of systems biology. I also would like to thank all his group members, especially Dr. Ayesha Shajahan, Dr. Harini Aiyer, Dr. Caroline Facey, Alan Zwart, Dr. Anni Warri, Elli Balakhanlou, Diane Demas, Ahreej Eltayeb, Dr. Katherine Cook, and Jessica Schwartz.

I would like to thank my Ph.D. advisory committee members including Dr. William T Baumann, Dr. Carla Finkielstein, and Dr. Liwu Li. Dr. Li really believed in my work and he really supported my ideas on my project.

I would like to thank Dr. Ran Kafri (Department of Systems Biology at Harvard Medical School). Ran was the first person who encouraged me to start a Ph.D. in systems biology and he was my first teacher on fluorescence microscopy. Ran was/is/will be an inspirational scientist for me, and besides being a great scientist, he is a great human being.

I would like to thank Dr. Yangqing Xu (Department of Systems Biology at Harvard Medical School). Yangqing kindly shared his experiences on live imaging of autophagy with me and helped me on my project a lot.

I would like to thank Dr. Galit Lahav (Department of Systems Biology at Harvard Medical School). I spent about 2 months in her lab and she and her group members kindly helped me to learn quantitative microscopy on cancer cells.

I would like to thank my family and my wife (Elnaz) who made it possible for me to pursue my Ph.D.

Dr. Mohsen Gheisarieha has been a great friend to me, and during these years, he has helped me a lot and I would like to thank him for being an exceptional friend. I also would like to thank his wife, Dr. Nancy H. Abdel-Wahab, because of her help and kind support.

I would like to thank Sarah Greene, Dr. Pizzi, and Buddy. I had a great time living with them during the last year of my Ph.D.

Madelon Carlisle, Nora, and Dylan have been great friends to me. Madelon has been an inspirational person for me, encouraging me to continue working on cancer and become an oncologist. I also should appreciate Zina Greene because of her kindness to me and our scientific discussions on cancer.

I would also like to thank Azadeh Aryan, Parsa, and Jalil Mirlohi for their kind help. Without Parsa's help, I could not finish writing this dissertation.

There have been a lot of great scientists who I have learned from their works or advices during my Ph.D., including Dr. Harold Garner (Virginia Tech), Dr. Hamid Bolouri (Fred Hutchinson Cancer Research Center), Dr. L. Samuel Wann (Wisconsin Heart Hospital in Milwaukee), Dr. Ilya Shmulevich (Institute for Systems Biology), Dr. David M. Moore (Virginia Tech), Dr. T.M. Murali (Virginia Tech), Dr. Omar I. Abdel-Wahab (Memorial Sloan-Kettering Cancer Center), Dr. Nima Mosammaparast (Washington University in St. Louis), Dr. Richard A. Lockshin (St. John's University), and Dr. Orkideh Behrouzan (King's College London). I should appreciate all of them.

I would like to thank my friends at Virginia Tech including Dr. Alireza Karimi, Dr. Arash Bahrami, and Abdullah Awaysheh.

Microscopic studies of autophagy in cancer cells were done at Microscopy and Imaging Shared Resource of Georgetown Lombardi comprehensive Cancer Center. For this part of my Ph.D. project, I need to appreciate Peter Johnson.

I would like to thank the program of Genetics, Bioinformatics and Computational Biology (GBCB) at Virginia Tech, especially Dennie Munson and Dr. David Bevan. I also would like to thank Virginia Bioinformatics Institute for awarding me the Transdisciplinary Team Science fellowship and Virginia Tech Graduate School for selecting me as 2011/2012 outstanding interdisciplinary doctoral student of Virginia Tech.

# Contents

**Iman Tavassoly (M.D., Ph.D.)**

Postdoctoral Research Fellow
Department of Pharmacology and Systems Therapeutics
Icahn School of Medicine at Mount Sinai
1425 Madison Avenue,
New York, NY10029

Email: iman.tavassoly@mssm.edu
Phone: 540-250-0930

## Education

| | | |
|---|---|---|
| Ph.D. (Genetics, Bioinformatics and Computational Biology), Outstanding Interdisciplinary Doctoral Program Student of Virginia Tech in 2011/2012 (Mentor: John Tyson) | August 2013 | Virginia Tech and Virginia Bioinformatics Institute |
| Graduate Certificate - Future Professoriate Certificate | Spring 2013 | Virginia Tech |
| M.D. | 2007 | Mazandaran University of Medical Sciences |
| Diploma (Biological Sciences and Mathematics) | 1999 | National Organization for Development of the Exceptional Talents (NODET), Iran. |

## Memberships

New York Academy of Sciences (2013–Present)
Virginia Tech Institutional Review Board (IRB) (2010–2013)
Interdisciplinary Research Honor Society at Virginia Tech (2010-Present)

## Research Interests

Systems Biology of Cancer, Cancer Genomics, Cell Death Pathways in Cancer, Personalized Medicine

## Research and Training Experiences and Appointments

| | | |
|---|---|---|
| High Performance Computing in the Life/Medical Sciences Summer Institute | Virginia Bioinformatics Institute | July/August 2014 |
| Howard-Hughes Medical Institute Workshop of "Teaching Systems Biology" | Center for Complex Biological Systems, UC Irvine | Jan 2013 |
| National Short Course in Systems Biology (Fellowship Award) | Center for Complex Biological Systems, UC Irvine | Jan 2013 |
| Visiting Scholar in "Physics and Mathematics of Cancer Program" (Fellowship Award) | The Kavli Institute for Theoretical Physics (KITP), UC Santa Barbara | May-July 2012 |
| Visiting Researcher and Graduate Student, (Project: Quantitative Fluorescence Microscopy of Autophagic Response in Cancer Cells) | Clarke Labs, Lombardi Comprehensive Cancer Center, Georgetown University Medical Centre | May 2011–Jan 2012 |
| Quantitative Fluorescence Microscopy Course Laboratory (Fellowship Award) | Mount Desert Island Biological Lab | May 2011 |
| NIH Sponsored PhosphoFlow/Immune Monitoring Training Course | Stanford University | 2011 |
| Computational Cell Biology Meeting | Cold Spring Harbor Laboratory | 2011 |
| Pharmaceutical Bioinformatics Course | Uppsala University, Sweden | 2011 |
| Visiting Graduate Student | Department of Systems Biology, Harvard Medical School | Dec 2010–Jan 2011 |
| Bootcamp in Cancer Modeling | Mathematical Biosciences Institute, Ohio State University | Sept 2010 |
| Visiting Graduate Student and Researcher | Clarke Labs, Lombardi Comprehensive Cancer Center | May 2010–Aug 2010 |

| | | |
|---|---|---|
| Workshop of "Dynamics of Signal Transduction and of Gene-Protein Regulatory Networks" | Mathematical Biosciences Institute, Ohio State University | Nov 2009 |
| International Summer School of Oncology (Fellowship Award) | Groningen University Medical Centre, the Netherlands. | 2006 |
| ISCOMS Research Fellow (Fellowship Award) | Groningen University Medical Centre, the Netherlands. | 2005 |
| Organizer of the Workshop: "Chaos and Fractals in Biologic Systems and Their Applications in Biomedical Sciences" | Groningen University Medical Centre, the Netherlands. | 2005 |

## Academic Awards and Honors

| | |
|---|---|
| Outstanding Interdisciplinary Doctoral Program Student Award of 2011/2012 | Virginia Tech |
| Best Poster Award in Virginia Tech 1st Cancer Symposium, 2011 | Virginia Tech |
| Transdisciplinary Team Science (TTS) Fellowship (Two-year fellowship award) 2009 | Virginia Bioinformatics Institute |
| Second Global Ubaydli Scholarship for Mobile Medical Computing, 2007 | USA |
| Outstanding Medical Student Researcher, 2004–2006 | Mazandaran University of Medical Sciences |
| Outstanding Medical Student of the Year, 2003 | Mazandaran University of Medical Sciences |

## Travel Fellowship Awards

| | |
|---|---|
| National Science Foundation Fellowship Award for High Performance Computing in the Life/Medical Sciences Summer Institute, 2014 | Virginia Tech |
| Fellowship Award for NIMBioS Investigative Workshop of Systems and Synthetic Microbiology, 2013 | University of Tennessee |
| Fellowship Award for National Short Course in: Systems Biology, 2013 | UC Irvine |

| | |
|---|---|
| Fellowship Award for "Physics and Mathematics of Cancer" Program, 2012 | KITP, UC Santa Barbara |
| Fellowship award for Quantitative Fluorescence Microscopy Course, 2011 | Mount Desert Island Biological Laboratory |
| Fellowship award for Summer School of Oncology, 2006 | Groningen University, the Netherlands |
| Fellowship award for ISCOMS Research Program, 2005 | Groningen University, the Netherlands |
| Fellowship Award for the International Student Congress of Medical Sciences, 2005 | Groningen University, the Netherlands |

## Publications

***Tavassoly I.***, Shajahan, A. N., Parmar, J., Baumann, W. T., Clarke, R., & Tyson, J. J.. "Dynamical modeling of the interaction between autophagy and apoptosis in mammalian cells: a systems pharmacology framework." arXiv preprint arXiv:1312.7149 (2013).

(Under Revision, CPT: Pharmacometrics and Systems Pharmacology, Nature Publishing Group)

Parmar J, Cook KL, Shajahan AN, Clarke PAG, ***Tavassoly I***, Clarke R, Tyson JJ, and Baumann WT, **Modeling the Effect of GRP78 on Anti-estrogen Sensitivity and Resistance in Breast Cancer**, ***Interface Focus*** 3 (4)

Clarke R, Cook KL, Hu R, Facey CO, ***Tavassoly I***, Schwartz JL et al. **Endoplasmic reticulum stress, the unfolded protein response, autophagy, and the integrated regulation of breast cancer cell** fate. ***Cancer Research*** 2012; 72(6): 1321–31.

Tyson JJ, Baumann WT, Chen C, Verdugo A, ***Tavassoly I,*** Wang Y et al. **Dynamic modelling of oestrogen signalling and cell fate in breast cancer cells.** ***Nature Reviews Cancer*** 2011; 11(7): 523–32.

Clarke R, Shajahan AN, Wang Y, Tyson JJ, Riggins RB, Weiner LM, Baumann WT, Xuan J, Zhang B, Facey CO, Aiyer H, Cook KL, Hickman FE, ***Tavassoly I***, et al. **Endoplasmic reticulum stress, the unfolded protein response, and gene network modeling in antiestrogen resistant breast cancer.**

Hormone ***Molecular Biology and Clinical Investigation*** 2011; 5(1): 35–44.

***Tavassoly I,*** Tavassoly O, Rad M.S.R, Dastjerdi N.M, **Multifractal Analysis of Chaos Game Representation Images of mtDNA**, ***Proceeding of the IEEE Conference*** Frontiers in the Convergence of Bioscience and Information Technologies (FBIT 2007), October 2007, pp 224–229, IEEE Press, ISBN 0-7695-2999-2.

***Tavassoly I,*** Tavassoly O, Rad M.S.R, Dastjerdi N.M, **Three Dimensional Chaos Game Representation of Genomic Sequences**, ***Proceeding of the IEEE***

***Conference*** Frontiers in the Convergence of Bioscience and Information Technologies (FBIT 2007), October 2007, pp 219–223, IEEE Press, ISBN 0-7695-2999-2.

***Tavassoly I,*** Tavassoly O, Rad M.S.R, **Chaos Game Representation of mtDNA: Is it Useful in Phylogenetic Studies?** (Abstract), ***BMC Systems Biology*** 2007, 1(Suppl 1):P35.

***Tavassoly I,*** Tavassoly O, **Fractal Analysis of Radiological Images: A Novel Quantitative Approach** (Abstract), ***Iranian Journal of Radiology*** (2007), 4 (Supp. 2), p: 24.

Tavassoly O, Namin S.A.M, Akbari M.T., ***Tavassoly I,*** **Utilization of Cell-Free Fetal DNA in Maternal Plasma for Prenatal Diagnosis of Paternity as a Non-invasive Method** (Abstract), The ***Journal of Maternal-Fetal & Neonatal Medicine*** (2008), 21 (Supp.1), p:284.

## Book Contributions

***Gene Therapy*** in "Encyclopedia of Cancer and Society" (Colditz G.A), SAGE publications, 2008

***Gene Silencing*** in "Encyclopedia of Global Health" (Zhang Y), SAGE publications, 2008

***Gene Array Analysis*** in "Encyclopedia of Global Health" (Zhang Y), SAGE publications, 2008.

***Hormone Disorders*** in "Encyclopedia of Obesity" (Keller K), Sage Publications 2008

***Histamines*** in "Encyclopedia of Obesity" (Keller K), Sage Publications 2008

***Mendelian Disorders Related to Obesity*** in "Encyclopedia of Obesity" (Keller K), SAGE Publications 2008

***CD36 and FAT*** in "Encyclopedia of Obesity" (Keller K), Sage Publications 2008

## Talks in Scientific and Medical Meetings

***Dynamical Modeling of Cell Death Pathways in Cancer: A Systems Pharmacology Framework***(Invited)

*Biomedical Computing Interest Group, National Institute of Health, September 2014.*

***Systems Biology of Cancer: Methods and Clinical Implications*** (Invited)

*Groningen University Medical Center, Groningen, the Netherlands, July 2014*

***Autophagic Response in Cancer Cells: Mathematical Modeling and Experimental Observations***(Invited)

*Kavli Institute of Theoretical Physics (KITP), UC Santa Barbara,* June 2012

***Autophagic Response in Single Cancer Cells***

Virginia Tech GSA Research Symposium, March 2012

***Systems Biology of Cell Death Pathways in Breast Cancer Cells***
National Cancer Institute's ICBP (Integrative Cancer Biology Program) Junior Investigators Meeting, MIT 2011

***Mathematical Modeling of Cell Fate Decision Mediated by Autophagy in Breast Cancer Cells***
Georgetown Lombardi Cancer Center, June 2011

***Multifractal Analysis of Chaos Game Representation Images of mtDNA*,**
Workshop of Emerging Topics in Human Functional Genomics and Proteomics, Turkey 2006

***Screening of the Patients with Arrhythmia from Healthy Subjects Based on Fractal Dimension of Heart Rate Variability by an Artificial Neural Network***
$12^{th}$ International Students Congress of Medical Sciences, (ISCOMS), Groningen University Medical Center, the Netherlands, 2005

## Poster Presentations at Scientific Meetings

*Dynamical Modeling of Cell Fate Decision Mediated by Autophagy in Breast Cancer Cells*
Southern California Systems Biology Conference, UC Irvine, 2013
Virginia Bioinformatics Institute Research Day, 2012
Computational Cell Biology Meeting, Cold Spring Harbor Laboratory, 2011
Virginia Tech Cancer Research Symposium, Virginia Tech 2011 (Best Poster Award)

*Mathematical Modeling of Autophagy and Unfolded Protein Response (UPR) Interplay in Breast Cancer Cells*
National Cancer Institute's ICBP (Integrative Cancer Biology Program) Mathematical Biology Meeting, UC Berkeley 2011

*Utilization of cell-free fetal DNA in Maternal Plasma for prenatal diagnosis of paternity as a non-invasive method*
XXI European Congress of Perinatal Medicine, Turkey 2008

*Multifractal Analysis of Chaos Game Representation Images of mtDNA*
Systems Biology, Bioinformatics and Synthetic Biology Forum, Manchester University, 2007

*Esophageal Cancer Belt of South-Central Asia*
International Summer School Oncology for Medical Students, Groningen University Medical Center, the Netherlands, 2006

## Teaching Activities

**Medical Sciences Summer School Oncology Groningen, The Netherlands** [June 2014–July 2014]
Lecturer in research track and Systems Biology of Cancer

**Coursera.org and Icahn School of Medicine at Mount Sinai** [March 2014 – June 2014]:

Teaching assistant in "Dynamical Modeling Methods for Systems Biology". (As a part of Systems Biology Specialization on Coursera)

Prepared content for quizzes, exams and online discussion forums for the massive open online course (MOOC) run by Coursera. Actively participated in the administration of the course; monitored and moderated discussion forums. (27356 students with 10116 active participants and over 1216 active discussion threads)

Further details of the course, including lectures and exams, are available online at https://www.coursera.org/specialization/systemsbiology/6?utm_medium=courseDescripTop

### *Virginia Tech*

Graduate Undergraduate Mentoring Program (GUMP), Virginia Tech, 2012 (Mentoring 5 Undergraduate Students on Systems Biology of Cancer)

K-12 Science Teaching (Kids' Tech University), Virginia Bioinformatics Institute, 2011. (Hands- On Exhibits of DNA Model and Heart Sounds)

## Extracurricular Activities

Invited Speaker at Virginia Tech Interdisciplinary Research Day, 2012

Published Photos: New York Times Photo Blog (2011) and New England Journal of Medicine (2011), Annals of Internal Medicine (2014)

Photography awards: Virginia Tech Graduate School Photo Contest Prize, Virginia Tech, 2009 and 2010, National Art Program Award (Mount Sinai Health System, New York, 2013)

Speaker at April $16^{th}$ Memorial Day (Peace and Poetry),Virginia Tech

Diploma in Independent Film-Making and Photography, Iranian Young Cinema Society,

2004 Director of a Short Film, Iranian Young Cinema Society, 2004

## Language Abilities

Persian (Native Language)

Fluent in English

German (Basic)

# Chapter 1
# Introduction to Autophagy in Physiology and Pathophysiology

Macroautophagy (referred to hereafter as autophagy) is a conserved catabolic cellular process by which a cell degrades its own components including damaged proteins and organelles. Autophagy is characterized by formation of autophagosomes, which are subcellular organelles enclosed by two or more membranes. Autophagosomes engulf damaged materials and degrade them. Autophagosomes dock with lysosomes, and the resulting autolysosome uses lysosomal enzymes to degrade the contents of the autophagic vacuoles [26, 87].

The molecular pathways taking part in autophagosome formation (Fig. 1.1) are known to some extent in yeast and mammalian cells. The main components are autophagy-related proteins (ATG proteins). The earliest steps of autophagosome formation are the induction and formation of an isolation membrane (phagophore), a small flattened membrane sac that elongates and curves to make an autophagosome [87]. Proteomic analysis has shed light on the global protein interaction network controlling autophagy [9].

Autophagy functions as a major component in mammalian cell homeostasis and is involved in pathogenesis of many diseases. Autophagy has some important roles in physiology and pathophysiology of plant cells as well [7, 8]. Autophagy is a part of starch degradation pathway in plant cells [133] and negatively controls the hypersensitive response programmed cell death during innate immunity response in plant cells, which allows the restriction of the hypersensitive response to the infected site [84]. Autophagy also is a protective stress response in plant cells, for instance, it is responsible for degrading oxidized proteins during oxidative stress in plants [141, 142].

While autophagy is normally initiated as a pro-survival stress response, excessive stress can trigger cell death [26, 87]. For example, in response to a lack of energy in the cell, protein folding in the endoplasmic reticulum (ER) is compromised and the

 
I. Tavassoly, *Dynamics of Cell Fate Decision Mediated by the Interplay of Autophagy and Apoptosis in Cancer Cells*, Springer Theses, DOI 10.1007/978-3-319-14962-2_1

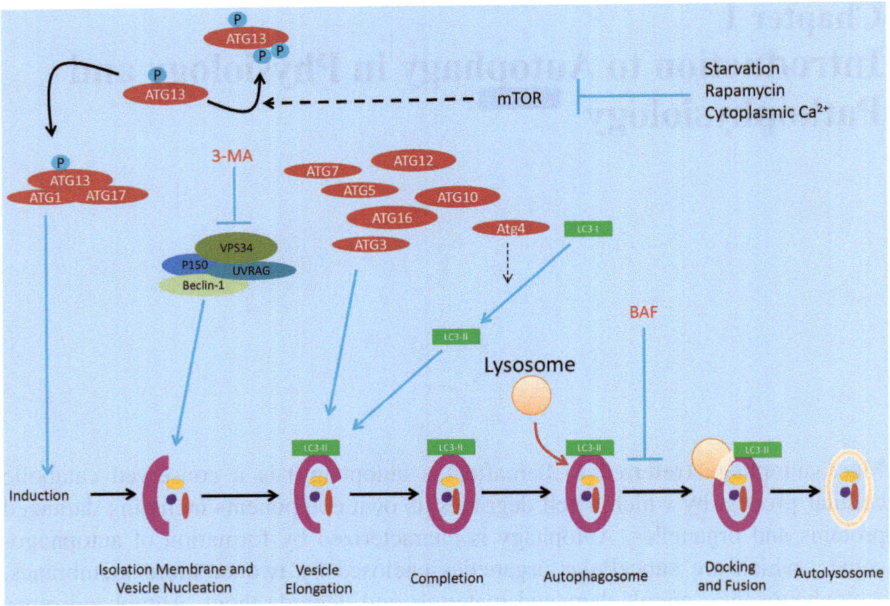


**Fig. 1.1** Different stages of autophagy and signaling molecules involved in them

unfolded protein response (UPR) is initiated [25]. Signals from the UPR up-regulate autophagy in an attempt to recover ATP and raw materials, to alleviate the stress [127, 130].

The autophagic response is not switch-like but gradual, with the level of autophagy increasing with the level of stress, analogous to a dimmer switch or rheostat rather than a toggle switch [127, 143]. Prolonged stress, which cannot be resolved by autophagy, will ultimately trigger cell death, often via apoptosis. Unlike autophagy, apoptosis is a switch-like process from which there is no return. Multiple cell death pathways have been reported in mammalian cells including apoptosis, autophagic cell death, necrosis, entosis, and autoschizis [35, 39, 42, 57, 88, 100, 127].

## 1.1 Cell Death Modalities in Cancer

Apoptosis has been studied extensively. Since its discovery in 1842 by Carl Vogt [131], mechanisms controlling this form of programmed cell death have been identified in different cell types including cancer cells. Apoptosis (Type I Programmed Cell Death) is morphologically characterized by cell shrinkage, loss of mitochondrial membrane potential, plasma membrane blebbing, and nuclear fragmentation [52]. The intrinsic pathway of apoptosis involves mitochondrial outer membrane permeabilization (MOMP) and cytochrome c release followed by caspases activation [78, 87]. Extrinsic pathway of apoptosis, which leads to the caspase-dependent apoptosis, is induced by activation of death receptors such as CD95/Fas and TRAIL [78, 87].

Autophagic cell death (Type II Programmed Cell Death) is characterized by massive cytoplasmic vacuolization, loss of multiple organelles, and the appearance of cytoplasmic subcellular organelles called autophagosomes [52, 76]. Autophagic cell death is a morphologically described cell death when autophagosomes are present in cytoplasm and cell death happens. Whether it is a new form of cell death (cell death by autophagy) or it is apoptosis happening after autophagic activity (cell death with autophagy) is still a controversy [76]. In Chap. 3 we discuss the cell morphologies which may have led to using term "Autophagic Cell Death."

Necrosis (Type III Programmed Cell Death) is defined morphologically in terms of progressive cell and organelle swelling and plasma membrane rupture [52]. Necrosis has been known to be an accidental and passive form of cell death that is not regulated by signaling pathways. But recently a programmed form of necrosis (necroptosis) has been found in different types of cells. Necroptosis is mediated by TNFalpha, and, although the details of its molecular regulatory network have poorly understood, it has been recognized as a tightly regulated form of stress response and cell death [23, 34, 38, 128]. Necroptosis is induced when apoptosis and autophagy pathways are blocked [139].

Autoschizis is a form of cell death which has been reported in different types of cancer cells. Autoschizis has the morphological features of both apoptosis and necrosis and is induced by treating cancer cells with vitamin C (VitC), vitamin K3 (VitK3), or a combination of both [11, 39, 40, 57]. Autoschizic cell death seems to be an atypical form of necrosis characterized by little condensation of heterochromatin, DNA intercalation of Vitamin K3, formation of autoschizis bodies, cell rounding, decrease in cell volume, and nuclear rounding [57]. Autoschizis bodies are results of cytoplasmic self-morsellation. During this self-morsellation, some parts of cytoplasm are excised while the excised parts do not contain any organelles [57]. It has been suggested that combination of vitamins with chemotherapy agents can be considered for a better response in cancer treatment via inducing autoschizic cell death [11, 57].

Anoikis is a form of apoptosis induced when epithelial cells are detached from the extra-cellular matrix (ECM) [36, 37]. Overholtzer et al. [100] have reported a new form of cell death observed in MCF10 and MCF7 cells when they are detached from ECM. This form of cell death is called "entosis" and is characterized by cell-in-cell structures [100]. Entosis happens when a detached epithelial cell enters another epithelial cell. The inside cell can have different fates including cell death (degradation by lysosomal enzymes of the host cell), mitosis or even it can be released [100]. Recently Florey et al. [35] have detected autophagic activity during entosis. They have demonstrated that autophagy can contribute to degradation of the internalized cell during entosis [35]. Entosis also acts as an inducer of aneuploidy in cancer cells [59]. Entosis is included among the "cell cannibalism" morphologies which are found in tumors. Cannibalistic cells are more common in malignant tissues compared to benign neoplasias [2, 44, 117].

## 1.2 Autophagy Pathway

Autophagy is the process of formation of autophagosomes, their docking with lysosomes and degradation. In fact autophagy is a multistep phenomenon which is mainly controlled by ATG proteins coded by *ATG* genes [87]. The very early step of autophagosome formation is induction of autophagy controlled by ATG13, ATG1, and ATG17. Autophagy induction needs a signal from mTOR which begins from dephosphorylation of ATG13. Formation of isolation membrane and vesicle nucleation is controlled by Beclin-1 (ATG6) complex and can be inhibited by some drugs such as 3-Methyladenine (3-MA). Then the vesicle engulfing cytoplasmic components elongates through action of ATG7, ATG12, ATG5, ATG16, ATG10, ATG3, and ATG4. In this process ATG8 (LC3) is anchored in the autophagosome membrane [87].

After assembly of autophagosomes, they dock with lysosomes. Drugs such as bafilomycin (BAF) can inhibit this step. Within autolysosomes, autophagosomes and their contents are degraded by lysosomal enzymes [87].

The main steps committing a cell to autophagy seem to be at the earliest stages of vesicle nucleation and formation of the isolation membrane [77, 87]. Major molecular players in the induction of autophagy in mammalian cells are mTOR (the mammalian target of rapamycin) and ATG13. mTOR is a signal integrator that senses stress conditions such as ER stress, hypoxia, low growth factor levels, or low levels of essential amino acids [61, 77, 87, 103, 109].

When there is no critical stress condition in the cell, mTOR represses a protein complex consisting of ULK1 (the mammalian homolog of Atg1), mammalian ATG13, and focal adhesion kinase interacting protein of 200 kD (FIP200) by phosphorylating ULK1 and ATG13 (Fig. 1.2). Cellular stress inactivates mTOR, allowing ULK1 and ATG13 to be dephosphorylated. ULK1 is then phosphorylated at a different site followed by auto-phosphorylation to become active. Active ULK1 phosphorylates ATG13 and FIP200 making them active which produces the active ULK1/ATG13/FIP200 complex. The active complex promotes formation of the isolation membrane [20, 61, 116].

*Beclin-1*, the mammalian ortholog of the yeast gene *Atg6*, is necessary for autophagosome formation, playing a key role in vesicle nucleation [50, 87]. Decreased *Beclin-1* expression or activity is associated with increased susceptibility to develop cancer [50]. BCL2 family proteins in the ER function as anti-autophagy proteins through their inhibitory interaction with Beclin-1. Although Beclin-1 is a BH3-only protein, it is not pro-apoptotic [50]. For autophagosome formation to begin, Beclin-1 must be released from BCL2 inhibition, which is promoted by either phosphorylation of BCL2 by c-Jun N-terminal kinase (JNK) or by phosphorylation

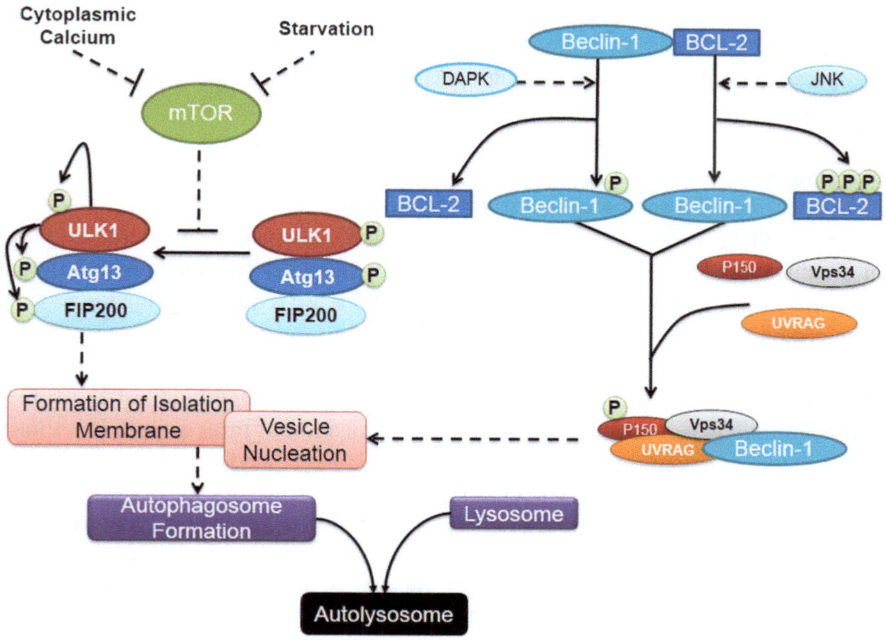


**Fig. 1.2** Formation of autophagosomes in response to stress

of Beclin-1 by death-associated protein kinase (DAPK) (Fig. 1.2) [62, 91, 136, 149, 150]. Free Beclin-1 can form a Beclin-1 core complex by binding to UVRAG (UV irradiation resistance-associated tumor suppressor gene) and P150 (a myristylated kinase); this complex can activate Vps34, a class III phosphatidylinositol 3-kinase (PI3K) [50, 77, 87]. The Beclin-1 core complex promotes vesicle nucleation [87].

A number of methods have been used for quantitative measurement of autophagosome formation in mammalian cells. One particularly convenient measure is the state of microtubule-associated protein light chain 3 (LC3), which is a mammalian homolog of yeast Atg8. LC3 exists in two forms: LC3-I (cytosolic form) and LC3-II (membrane-bound form). After autophagy initiation, LC3-I is converted to LC3-II, which then participates in the vesicle elongation step of autophagosome formation [71, 87].

## 1.3 Selective Autophagy

Autophagy was considered to be a non-selective bulk process, but recently several forms of selective autophagy have been found [115, 149].

Regardless of being selective or non-selective, three different types of autophagy exist and usually the term "autophagy" refers to "Macroautophagy." Other types of autophagy include Chaperone-Mediated Autophagy (CMA) and Microautophagy.

Microautophagy is the process of degrading cytoplasmic components by lysosomes when lysosomes directly engulf these components and digest them. Microautophagy can be non-selective while selective forms of microautophagy have been reported including micropexophagy (microautophagy of peroxisomes), micromitophagy (microautophagy of mitochondria), and piecemeal microautophagy of the nucleus [74, 83, 149]. The signaling pathway of microautophagy is different from macroautophagy but some of the ATG proteins are required during the process of microautophagy [74].

CMV is a form of autophagy in which unfolded and misfolded cytoplasmic proteins are targeted and directly translocated across the membrane of lysosomes and degraded by lysosomal enzymes. The translocation of proteins into lysosomes is mediated by chaperone HSC70 and lysosome-associated membrane protein 2 (LAMP2). CMV is a selective process [31, 91].

Macroautophagy is the process of engulfing cytoplasmic materials by formation of autophagosomes. Macroautophagy can be a bulk and non-selective process. Selective forms of macroautophagy include mitophagy, pexophagy, aggerephagy, glycophagy, ribophagy, lipophagy, ER-phagy, xenophagy, allophagy, and zymophagy [13, 22, 33, 43, 60, 67, 99, 123, 144].

Mitophagy is selective degradation of damaged mitochondria by autophagy. Mitophagy acts as a quality control system for mitochondria to keep the cell energy homeostasis and to prevent accumulation of reactive oxygen species (ROS) produced by mitochondria [67, 96, 126, 146]. Knockdown of mitophagy results in dysfunctional and ROS-generating mitochondria [151].

ER-phagy is selective degradation of endoplasmic reticulum by autophagosome formation [33, 147]. ER-phagy prevents excessive expansion of endoplasmic reticulum during ER stress [12]. Bernales et al. [12] have observed the expansion of endoplasmic reticulum volume after ER stress and induction of UPR in yeast cells and simultaneous formation of autophagosomes attached to expanded ER which are responsible to selectively degrade the ER.

Ribosomes can be degraded non-selectively under physiological conditions, but upon stress they can be degraded by a form of autophagy called "Ribophagy." Ribophagy contributes to conservation of ATP and amino acid because degradation of ribosomes not only provides amino acids by itself but also reduces the protein translation [60, 99].

Glycophagy and lipophagy are selective forms of macroautophagy for degradation of glycogen granules and lipid droplets, respectively [60, 137], and zymophagy is macroautophagic degradation of secretory granules [43, 60].

Xenophagy is a selective form of macroautophagy that plays role as an immune mechanism to target and degrade foreign pathogens such as bacteria and viruses [60, 123].

Allophagy is a form of macroautophagy designed to degrade the paternal organelles coming from spermatozoid during fertilization [3].

Recently a genome-wide siRNA screen has found mammalian genes involved in selective autophagy [98]. Quantitative mass spectrometry of temporal degradation of protein during starvation-induced autophagy has shown that autophagy degrades proteins and organelles in a specific order [75]. Kristensen et al. [75] showed that protein degradation by autophagy starts from cytosolic and proteosomal proteins and then proceeds to degradation of organelles (mitophagy, ribophagy, ER-phagy, etc.) [75].

As mentioned before, our work is focused on macroautophagy (referred to as autophagy).

## 1.4 Autophagy and Pathogenesis of Diseases

Autophagy and its dysregulation play important roles in the pathogenesis of many complex diseases [91, 149, 150]. Aging has been related to attenuated autophagy, and autophagy is known as an anti-aging mechanism [28, 125]. Autophagy is a cellular mechanism linking the beneficial metabolic effects of exercise and health. During exercise, autophagy is induced and controls the metabolism of muscles [51].

Since autophagy is an essential element for homeostasis in neurons, dysfunctions in its regulation can cause a range of neurodegenerative disorders [47, 72, 94]. Impaired basal autophagy in neurons leads to neurodegenerative changes [47]. These neurodegenerative changes are caused by aggregation of dysfunctional mitochondria and damaged proteins in cytoplasm, and neurons use autophagy as a clean-up system to avoid neurodegeneration [22]. Induction of autophagy by inhibitors of mTOR reduces neurodegenerative features in mice [111].

Increased autophagy is also defined within hallmarks of cancer cells, because they need to survive the stresses of nutrient deprivation and anoxia [46]. Autophagy has been reported as a process in connection with cell death pathways in cancer cells, and it has been shown that it is also a mechanism for elimination of cancer cells. *Beclin1* gene has reduced copy numbers in some cancer cells, and lack of one of the two copies in *Beclin1* gene is related to more tumor incidence [138]. In addition, autophagy is involved in the development of resistance to chemotherapies, and inhibiting autophagy can increase the therapeutic responses of resistant cancer cells to chemotherapy, endocrine therapy, and/or radiation therapy [4, 19, 24, 26, 27, 58, 73, 85, 102, 108]. Autophagy also takes part in the coupling of metabolic activities of cancer cells and tumor stromal cells [105]. Figure 1.3 shows the basal extent of autophagy in a human H4 neuroglioma cell expressing GFP-LC3 (Green dots are GFP-LC3 puncta).

Autophagy is a protective mechanism by which myocytes maintain normal homeostasis, and lack of autophagy is a dominant feature of cardiovascular diseases such as myocardial hypertrophy and cardiomyopathies. Autophagy is also a protective mechanism in ischemic heart diseases [45, 54, 93, 124, 145].

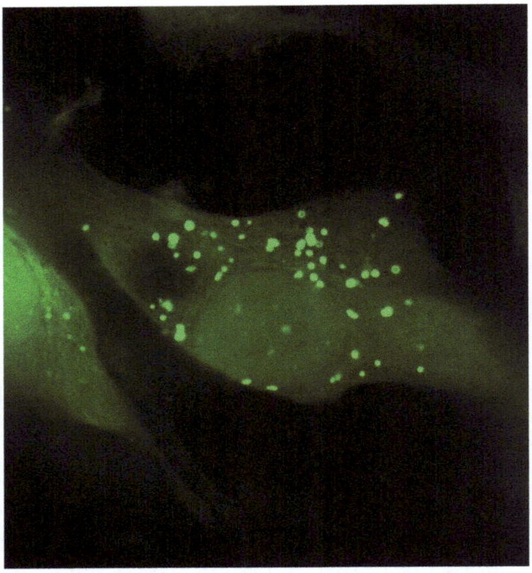

**Fig. 1.3** Basal autophagy in a single human H4 neuroglioma cell expressing GFP-LC3

*In vitro* and *in vivo* experiments have shown that autophagy is an important protective response in renal pathophysiology [55, 56, 65, 106, 122]. Cisplatin is a chemotherapy agent which causes acute kidney injury as a side effect. Autophagy is stimulated in renal tubular kidney cells as a protective mechanism to postpone the initiation of cell death after cisplatin treatment [65, 106].

Autophagy also acts as an immune response and is active in the elimination of intracellular pathogens such as bacteria and viruses, while some of these pathogens may hijack the autophagy pathway to facilitate their own replication in cells [30, 80, 112].

Because of the key role of autophagy in pathogenesis of many diseases, a detailed and integrative understanding of molecular and cellular mechanisms regulating it will help to target autophagy for therapeutic purposes.

## 1.5 Autophagy and Metabolism in Cancer

Autophagy is induced by a wide range of stresses such as starvation, hypoxia, ROS, drug treatments, DNA damage, and even mechanical stress [1, 6, 26, 64, 66, 70, 73, 77, 81, 90, 95, 101, 110].

Autophagy initially is a prosurvival mechanism to rescue cells from stress. Especially in context of cancer biology, autophagy contributes to blocking cell death pathways and to the development of therapeutic resistant phenotype in tumors. How autophagy can suppress stresses (such as drug treatments) and postpone cell death is probably through products of autophagic process such as ATP and nutrient sources which help cells to resist stresses. In addition, a lot of stresses cause unfolded proteins and dysfunctional organelles, and autophagy degrades them as a quality control process in cells. Some forms of autophagy like (mitophagy) reduces ROS in cells, and ribophagy decreases protein translation which helps to restore a high level of ATP and amino acids [60, 99]. Another function of autophagy in cell survival and stress attenuation can be its role in cancer cell metabolism. Autophagy is used in the interaction of cancer cells with their microenvironment.

Normal cells produce ATP mainly by oxidation of pyruvate in mitochondria. When oxygen is available they use glycolysis to convert glucose to pyruvate and after that pyruvate is oxidized in the mitochondria. During this process (glycolysis plus oxidative phosphorylation) approximately 36 mol ATP is produced per 1 mol glucose [79, 129].

Under oxygen deprivation, normal cells use anaerobic glycolysis, in which pyruvate is reduced to lactate so that glycolysis may continue. Anaerobic glycolysis produces 2 mol ATP per 1 mol glucose [79, 129].

Otto Warburg in 1956 [134] found that cancer cells use glycolysis as their main pathway of ATP production instead of oxidative phosphorylation. He observed that cancer cells convert glucose to lactate both when oxygen is available and when there is lack of oxygen [79, 129]. This phenomenon, which is called the "Warburg Effect" or aerobic glycolysis is not as efficient as oxidative phosphorylation in terms of ATP production because it produces about 4 mol ATP per mol glucose [79, 129, 134].

In 2009 Pavlides et al. [104] reported another interesting metabolic phenomenon in cancer cells which is called the "Reverse Warburg Effect." They observed that epithelial cancer cells cause the initiation of Warburg effect in their neighboring stromal fibroblasts, which are called cancer-associated fibroblasts. Products of aerobic glycolysis (pyruvate and lactate) are secreted by cancer-associated fibroblasts and are used by epithelial cancer cells for tricarboxylic acid cycle (TCA) in mitochondria and oxidative phosphorylation. The reverse Warburg effect lets cancer cells produce ATP in an efficient way [10, 104, 120]. Autophagy is one of the mechanisms used to control the Reverse Warburg Effect.

Cancer cells produce and release ROS, which activate both autophagy (mitophagy) and aerobic glycolysis in cancer-associated fibroblasts or tumor stromal cells [10, 119]. On the other hand, as cancer cells are addicted to glutamine [29, 140], metabolism of glutamine in cancer cells leads to ammonia production. Ammonia can induce autophagy in tumor stromal cells. Autophagy in tumor stromal cells provides new source of glutamine which can be used by cancer cells [119]. Glutamine (as a product of autophagy) and ketones and lactate (as products of aerobic glycolysis) in tumor stromal cells provide cancer cells with substrates for their efficient oxidative phosphorylation. The combination of reverse Warburg effect and autophagy in both cancer cells and tumor stromal cells brings a very good reservoir of ATP and nutrients for cancer cells to be used in resisting stress [10, 17, 119]. Figure 1.4 has summarized the interaction of autophagy in cancer cells and cancer-associated fibroblasts with reverse Warburg effect phenomenon.

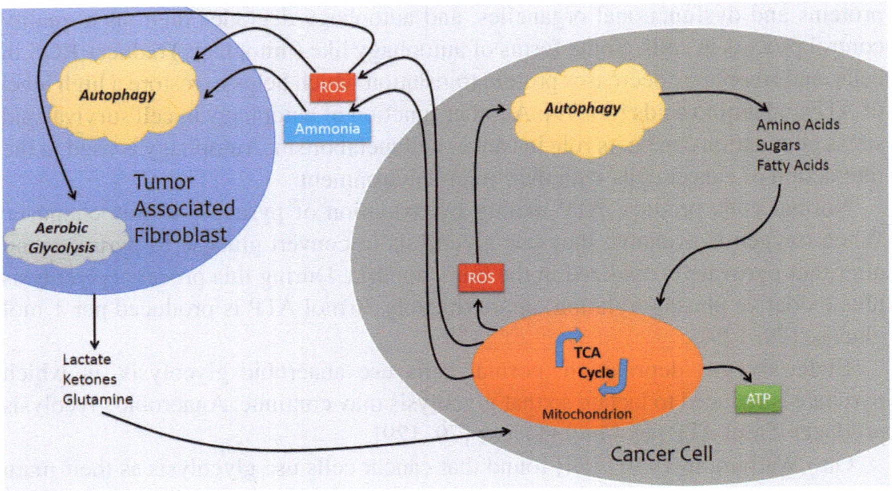


**Fig. 1.4** Autophagy in tumor cells can suppress the stress via interactions with metabolic pathways

## 1.6 Signaling Pathway Controlling Interplay of Autophagy and Apoptosis

The molecular regulatory pathway that links autophagy to apoptosis has been studied in various cell types. While there are still many unanswered questions regarding this pathway, it is evident that regulation can be accomplished by controlling calcium signaling from the endoplasmic reticulum to the mitochondria [14, 49, 68, 89, 97, 113].

Figure 1.5 is an influence diagram illustrating these pathways.

BCL2 family proteins in the ER and mitochondria are important regulators of autophagy and apoptosis, respectively [41, 114]. Hence, different levels of expression of BCL2 proteins in the ER membrane and the mitochondrial membrane can lead to different activation dynamics for autophagy and apoptosis in a variety of cell types. In addition, when stress is first induced, JNK phosphorylates BCL2 at the ER and affects only the BCL2/Beclin-1 complex [136]. Hence, autophagy is normally induced before apoptosis.

Figure 1.6 shows the details of the interactions among BCL2, Beclin-1, and inositol-1,4,5-trisphosphate receptor (IP3R). In this figure, solid arrows indicate chemical reactions (association, dissociation, phosphorylation, cleavage), and dashed arrows indicate catalytic activities.

The IP3R and the sarco/endoplasmic reticulum $Ca^{2+}$-ATPase (SERCA) pump are the central regulators of $Ca^{2+}$ exchange between ER and cytoplasm. By pumping $Ca^{2+}$ from the cytosol into the ER, SERCA is responsible for

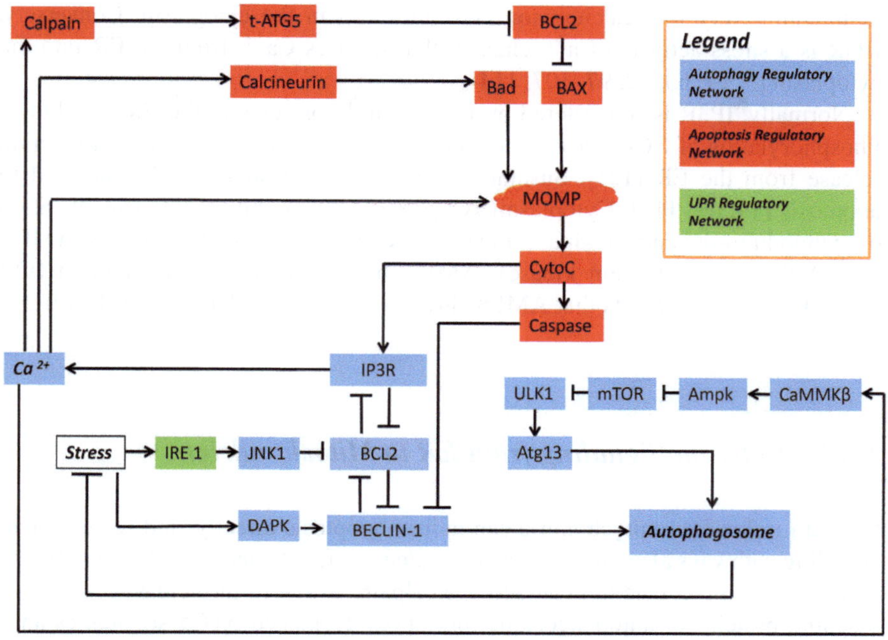


**Fig. 1.5** A detailed influence diagram for crosstalk of autophagy and apoptosis. *Arrowheads* indicate activation and *blunt heads* indicate inhibition

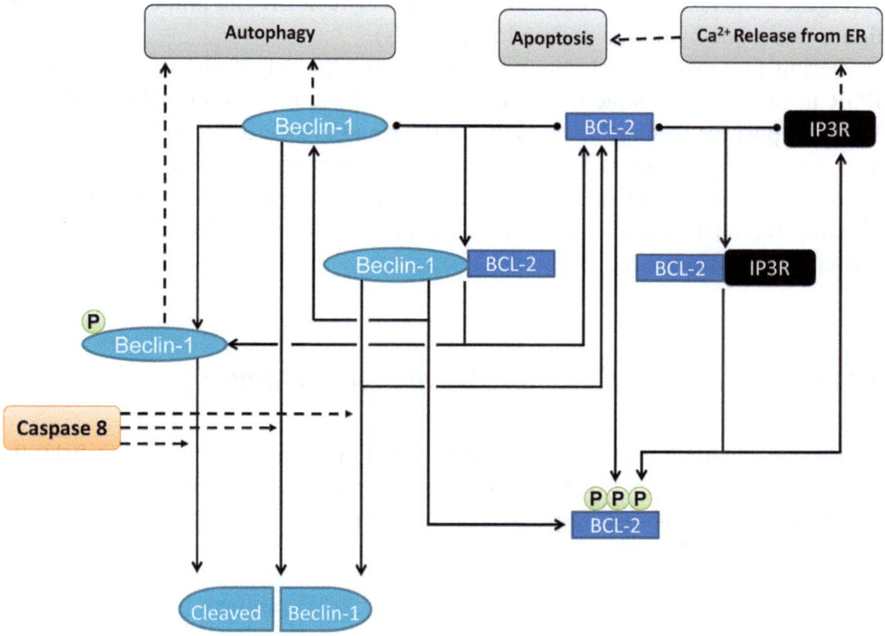


**Fig. 1.6** Interactions of anti-autophagy BCL2 family proteins with IP3R and Beclin-1 in the endoplasmic reticulum (ER)

maintaining very low calcium ion concentrations in the cytoplasm. Conversely, IP3R is a stress-activated $Ca^{2+}$ channel that releases $Ca^{2+}$ from the ER into the cytoplasm [5, 14, 63, 92, 97, 107, 114].

Normally, IP3R is sequestered by BCL2 family proteins in the ER membrane. Phosphorylation of BCL2 proteins dissociates the complex and allows for calcium release from the ER [121]. Sustained, elevated cytoplasmic $[Ca^{2+}]$ can lead to apoptosis [14, 49, 114]. Cytoplasmic $Ca^{2+}$ can also inhibit mTOR via activation of calmodulin-dependent kinase kinase-$\beta$ (CaMKK$\beta$), which activates AMPK (5′ AMP-activated protein kinase). AMPK has an inhibitory effect on mTOR [53]. Other work suggests that AMPK directly activates the ULK1/ATG13/FIP200 complex [48, 69].

### *1.6.1 Calcium Signaling from ER to Mitochondrion*

Calcium influx into mitochondria can induce apoptosis directly, and several other signaling pathways also link sustained elevated calcium to apoptosis [14, 18, 49, 89, 114]. For example, calcium can activate calpain, which in turn cleaves ATG5, an essential protein for autophagosome formation. Truncated ATG5 suppresses anti-apoptotic BCL2 proteins in mitochondria and induces apoptosis [86, 87, 148]. In addition, calcium activates calcineurin, which dephosphorylates and activates Bad, a pro-apoptotic BCL2 family protein capable of inducing apoptosis [68, 132].

Figure 1.7 illustrates the calcium signaling between ER and mitochondrion (this figure was made by online protein lounge tools).

During apoptosis, cytochrome c is released into the cytoplasm, where it binds to IP3R in the ER membrane to induce calcium release into the cytoplasm [15, 16]. Why this positive feedback loop between IP3R and cytochrome c exists in cells and how it affects apoptosis is an interesting but unanswered question. Cytochrome c release also leads to activation of caspases, which are inhibitors of autophagy. By cleaving Beclin-1, some caspases (e.g., caspase 8) down-regulate autophagosome formation [21, 32, 82].

### *1.6.2 DAPK Fine-Tunes the Autophagic Response*

DAPK is a signaling protein in both pathways of autophagy and apoptosis [91]. By phosphorylating Beclin-1, DAPK functions as a fine-tuner of calcium release from ER during the autophagic response.This action is done through role of DAPK in phosphorylation of Beclin-1 [91, 149, 150].

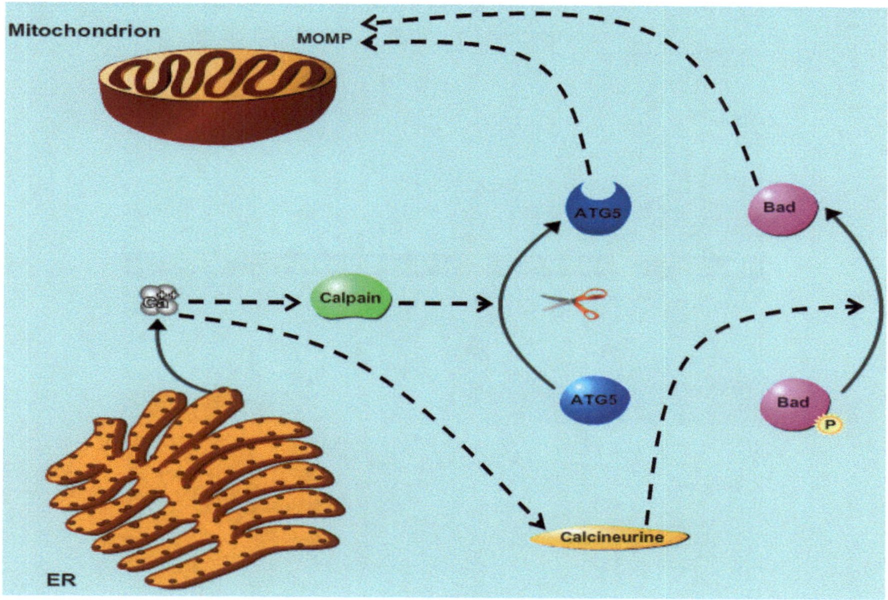


**Fig. 1.7** Transduction of stress signal from ER to mitochondrion through calcium signaling

Under physiological conditions (no stress), IP3R is in its inactive state because of BCL2 inhibition. Simultaneously BCL2 proteins in ER inhibit Beclin-1 [87, 121]. In this condition, there is no calcium release from IP3R and no autophagy (Fig. 1.8a).

When stress activates JNK, phosphorylation of BCL2 in ER has two major consequences: Calcium release from ER through IP3Rs and initiation of autophagy by free Beclin-1 [118, 135, 136] (Fig. 1.8b).

DAPK activation upon stress results in phosphorylation of Beclin-1 (Fig. 1.8c). Phosphorylated Beclin-1 can initiate autophagy. At the same some BCL2 proteins in ER are free to bind to and inactivate IP3R [91, 149, 150]. Hence, calcium release from ER via IP3R is reduced. This fine-tuning by DAPK provides a longer time delay between initiation of autophagy and induction of apoptosis (Figure 1.8 was made by online protein lounge tools).

In Chap. 2, we combine all this information into a single influence diagram for the interaction of autophagy and apoptosis. This diagram will be used as the foundation for our mathematical model.

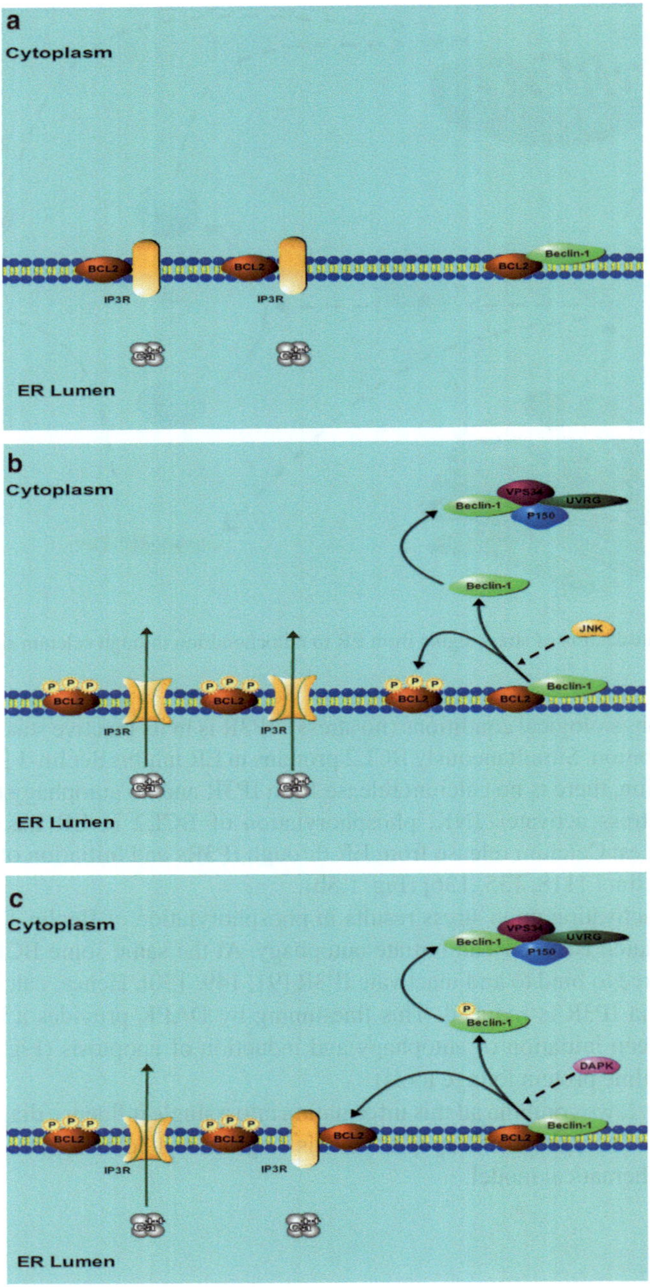


**Fig. 1.8** Interaction of BCL2, Beclin-1, and IP3R in ER

## References


1. Abedin, M.J., Wang, D., McDonnell, M.A., Lehmann, U., Kelekar, A.: Autophagy delays apoptotic death in breast cancer cells following DNA damage. Cell Death Differ. **14**(3), 500–510 (2006)
2. Abodief, W.T., Dey, P., Al-Hattab, O.: Cell cannibalism in ductal carcinoma of breast. Cytopathology **17**(5), 304–305 (2006)
3. Al Rawi, S., Louvet-Vallée, S., Djeddi, A., Sachse, M., Culetto, E., Hajjar, C., Boyd, L., Legouis, R., Galy, V.: Allophagy: a macroautophagic process degrading spermatozoid-inherited organelles. Autophagy **8**(3), 421–423 (2012)
4. Apel, A., Herr, I., Schwarz, H., Rodemann, H.P., Mayer, A.: Blocked autophagy sensitizes resistant carcinoma cells to radiation therapy. Cancer Res. **68**(5), 1485–1494 (2008)
5. Arbabian, A., Brouland, J.-P., Gélébart, P., Kovàcs, T., Bobe, R., Enouf, J., Papp, B.: Endoplasmic reticulum calcium pumps and cancer. Biofactors **37**(3), 139–149 (2011)
6. Azad, M.B., Chen, Y., Gibson, S.B.: Regulation of autophagy by reactive oxygen species (ros): implications for cancer progression and treatment. Antioxid. Redox Signal. **11**(4), 777–790 (2009)
7. Bassham, D.C.: Plant autophagymore than a starvation response. Curr. Opin. Plant Biol. **10**(6), 587–593 (2007)
8. Bassham, D.C., Laporte, M., Marty, F., Moriyasu, Y., Ohsumi, Y., Olsen, L.J., Yoshimoto, K.: Review autophagy in development and stress responses of plants. Autophagy **2**(1), 2–11 (2006)
9. Behrends, C., Sowa, M.E., Gygi, S.P., Harper, J.W.: Network organization of the human autophagy system. Nature **466**(7302), 68–76 (2010)
10. Bellot, G.L., Liu, D., Pervaiz, S.: Ros, autophagy, mitochondria and cancer: Ras, the hidden master? Mitochondrion **13**, 155–162 (2012)
11. Ben-Jacob, E., Coffey, D.S., Levine, H.: Bacterial survival strategies suggest rethinking cancer cooperativity. Trends Microbiol. **20**, 403–410 (2012)
12. Bernales, S., McDonald, K.L., Walter, P.: Autophagy counterbalances endoplasmic reticulum expansion during the unfolded protein response. PLoS Biol. **4**(12), e423 (2006)
13. Bernales, S., Schuck, S., Walter, P.: Er-phagy: selective autophagy of the endoplasmic reticulum. Autophagy **3**(3), 285–287 (2007)
14. Berridge, M.J., Bootman, M.D., Llewelyn Roderick, H.: Calcium signalling: dynamics, homeostasis and remodelling. Nat. Rev. Mol. Cell Biol. **4**(7), 517–529 (2003)
15. Boehning, D., Patterson, R.L., Sedaghat, L., Glebova, N.O., Kurosaki, T., Snyder, S.H.: Cytochrome c binds to inositol (1, 4, 5) trisphosphate receptors, amplifying calcium-dependent apoptosis. Nat. Cell Biol. **5**(12), 1051–1061 (2003)
16. Boehning, D., Patterson, R.L., Snyder, S.H.: Apoptosis and calcium: new roles for cytochrome c and inositol 1, 4, 5-trisphosphate. Cell Cycle **3**(3), 250–252 (2004)
17. Bonuccelli, G., Tsirigos, A., Whitaker-Menezes, D., Pavlides, S., Pestell, R.G., Chiavarina, B., Frank, P.G., Flomenberg, N., Howell, A., Martinez-Outschoorn, U.E., et al.: Ketones and lactate fuel tumor growth and metastasis: evidence that epithelial cancer cells use oxidative mitochondrial metabolism. Cell Cycle **9**(17), 3506–3514 (2010)
18. Cerella, C., Diederich, M., Ghibelli, L.: The dual role of calcium as messenger and stressor in cell damage, death, and survival. Int. J. Cell Biol. **2010**, 546163 (2010)
19. Chaachouay, H., Ohneseit, P., Toulany, M., Kehlbach, R., Multhoff, G., Rodemann, H.P.: Autophagy contributes to resistance of tumor cells to ionizing radiation. Radiother. Oncol. **99**(3), 287–292 (2011)
20. Chan, E.Y.: mtorc1 phosphorylates the ulk1-matg13-fip200 autophagy regulatory complex. Sci. Signal. **2**(84), pe51 (2009)
21. Cho, D.H., Jo, Y.K., Hwang, J.J., Lee, Y.M., Roh, S., Kim, J.C.: Caspase-mediated cleavage of atg6/beclin-1 links apoptosis to autophagy in hela cells. Cancer Lett. **274**(1), 95–100 (2009)

22. Choi, A.M.K., Ryter, S.W., Levine, B.: Autophagy in human health and disease. N. Engl. J. Med. **368**(7), 651–662 (2013)
23. Christofferson, D.E., Yuan, J.: Necroptosis as an alternative form of programmed cell death. Curr. Opin. Cell Biol. **22**(2), 263–268 (2010)
24. Clarke, R., Shajahan, A.N., Riggins, R.B., Cho, Y., Crawford, A., Xuan, J., Wang, Y., Zwart, A., Nehra, R., Liu, M.C.: Gene network signaling in hormone responsiveness modifies apoptosis and autophagy in breast cancer cells. J. Steroid Biochem. Mol. Biol. **114**(1), 8–20 (2009)
25. Clarke, R., Shajahan, A.N., Wang, Y., Tyson, J.J., Riggins, R.B., Weiner, L.M., Bauman, W.T., Xuan, J., Zhang, B., Facey, C., et al.: Endoplasmic reticulum stress, the unfolded protein response, and gene network modeling in antiestrogen resistant breast cancer. Horm. Mol. Biol. Clin. Invest. **5**(1), 35–44 (2011)
26. Clarke, R., Cook, K.L., Hu, R., Facey, C.O.B., Tavassoly, I., Schwartz, J.L., Baumann, W.T., Tyson, J.J., Xuan, J., Wang, Y., et al.: Endoplasmic reticulum stress, the unfolded protein response, autophagy, and the integrated regulation of breast cancer cell fate. Cancer Res. **72**(6), 1321–1331 (2012)
27. Cook, K.L., Shajahan, A.N., Clarke, R.: Autophagy and endocrine resistance in breast cancer. Expert Rev. Anticancer Ther. **11**(8), 1283–1294 (2011)
28. Cuervo, A.M., Bergamini, E., Brunk, U.T., Dröge, W., Ffrench, M., Terman, A.: Autophagy and aging: the importance of maintaining" clean" cells. Autophagy **1**(3), 131–140 (2005)
29. Dang, C.V.: Glutaminolysis: supplying carbon or nitrogen or both for cancer cells? Cell Cycle **9**(19), 3914–3916 (2010)
30. de Haan, C.A.M., Reggiori, F.: Are nidoviruses hijacking the autophagy machinery? Autophagy **4**(3), 276–279 (2008)
31. Dice, J.F.: Chaperone-mediated autophagy. Autophagy **3**(4), 295–299 (2007)
32. Djavaheri-Mergny, M., Maiuri, M.C., Kroemer, G.: Cross talk between apoptosis and autophagy by caspase-mediated cleavage of beclin 1. Oncogene **29**(12), 1717–1719 (2010)
33. Dunn, W.A. Jr., Cregg, J.M., Kiel, J.A., van der Klei, I.J., Oku, M., Sakai, Y., Sibirny, A.A., Stasyk, V., Veenhuis, M.: Pexophagy: the selective autophagy of peroxisomes. Autophagy **1**(2), 75–83 (2005)
34. Edinger, A.L., Thompson, C.B.: Death by design: apoptosis, necrosis and autophagy. Curr. Opin. Cell Biol. **16**(6), 663–669 (2004)
35. Florey, O., Kim, S.E., Sandoval, C.P., Haynes, C.M., Overholtzer, M.: Autophagy machinery mediates macroendocytic processing and entotic cell death by targeting single membranes. Nat. Cell Biol. **13**(11), 1335–1343 (2011)
36. Frisch, S.M., Francis, H.: Disruption of epithelial cell-matrix interactions induces apoptosis. J. Cell Biol. **124**(4), 619–626 (1994)
37. Frisch, S.M., Screaton, R.A.: Anoikis mechanisms. Curr. Opin. Cell Biol. **13**(5), 555–562 (2001)
38. Galluzzi, L., Kroemer, G.: Necroptosis: a specialized pathway of programmed necrosis. Cell **135**(7), 1161–1163 (2008)
39. Gilloteaux, J., Jamison, J.M., Arnold, D., Summers, J.L., et al.: Autoschizis: another cell death for cancer cells induced by oxidative stress. Archivio italiano di anatomia ed embriologia (Ital. J. Anat. Embryol.) **106**(2 Suppl. 1), 79 (2001)
40. Gilloteaux, J., Jamison, J.M., Arnold, D., Jarjoura, D., Von Greuningen, V., Summers, J.L.: Autoschizis of human ovarian carcinoma cells: scanning electron and light microscopy of a new cell death induced by sodium ascorbate: menadione treatment. Scanning **25**(3), 137–149 (2003)
41. Gotow, T., Shibata, M., Kanamori, S., Tokuno, O., Ohsawa, Y., Sato, N., Isahara, K., Yayoi, Y., Watanabe, T., Leterrier, J.F., et al.: Selective localization of bcl-2 to the inner mitochondrial and smooth endoplasmic reticulum membranes in mammalian cells. Cell Death Differ. **7**(7), 666 (2000)
42. Gozuacik, D., Kimchi, A.: Autophagy and cell death. Curr. Top. Dev. Biol. **78**, 217–245 (2007)

43. Grasso, D., Ropolo, A., Ré, A.L., Boggio, V., Molejón, M.I., Iovanna, J.L., Gonzalez, C.D., Urrutia, R., Vaccaro, M.I.: Zymophagy, a novel selective autophagy pathway mediated by vmp1-usp9x-p62, prevents pancreatic cell death. J. Biol. Chem. **286**(10), 8308–8324 (2011)
44. Gupta, K., Dey, P.: Cell cannibalism: diagnostic marker of malignancy. Diagn. Cytopathol. **28**(2), 86–87 (2003)
45. Gustafsson, A.B., Gottlieb, R.A.: Autophagy in ischemic heart disease. Circ. Res. **104**(2), 150–158 (2009)
46. Hanahan, D., Weinberg, R.A.: Hallmarks of cancer: the next generation. Cell **144**(5), 646–674 (2011)
47. Hara, T., Nakamura, K., Matsui, M., Yamamoto, A., Nakahara, Y., Suzuki-Migishima, R., Yokoyama, M., Mishima, K., Saito, I., Okano, H., et al.: Suppression of basal autophagy in neural cells causes neurodegenerative disease in mice. Nature **441**(7095), 885–889 (2006)
48. Hardie, D.G.: Ampk and autophagy get connected. EMBO J. **30**(4), 634 (2011)
49. Harr, M.W., Distelhorst, C.W.: Apoptosis and autophagy: decoding calcium signals that mediate life or death. Cold Spring Harb. Perspect. Biol. **2**(10), a005579 (2010)
50. He, C., Levine, B.: The beclin 1 interactome. Curr. Opin. Cell Biol. **22**(2), 140–149 (2010)
51. He, C., Bassik, M.C., Moresi, V., Sun, K., Wei, Y., Zou, Z., An, Z., Loh, J., Fisher, J., Sun, Q., et al.: Exercise-induced bcl2-regulated autophagy is required for muscle glucose homeostasis. Nature **481**(7382), 511–515 (2012)
52. Hotchkiss, R.S., Strasser, A., McDunn, J.E., Swanson, P.E.: Cell death. N. Engl. J. Med. **361**(16), 1570–1583 (2009)
53. Høyer-Hansen, M., Bastholm, L., Szyniarowski, P., Campanella, M., Szabadkai, G., Farkas, T., Bianchi, K., Fehrenbacher, N., Elling, F., Rizzuto, R., et al.: Control of macroautophagy by calcium, calmodulin-dependent kinase kinase-$\beta$, and bcl-2. Mol. Cell **25**(2), 193–205 (2007)
54. Hua, Y., Zhang, Y., Ceylan-Isik, A.F., Wold, L.E., Nunn, J.M., Ren, J.: Chronic akt activation accentuates aging-induced cardiac hypertrophy and myocardial contractile dysfunction: role of autophagy. Basic Res. Cardiol. **106**(6), 1173–1191 (2011)
55. Inoue, K., Kuwana, H., Shimamura, Y., Ogata, K., Taniguchi, Y., Kagawa, T., Horino, T., Takao, T., Morita, T., Sasaki, S., et al.: Cisplatin-induced macroautophagy occurs prior to apoptosis in proximal tubules in vivo. Clin Exp. Nephrol. **14**(2), 112–122 (2010)
56. Isaka, Y., Kimura, T., Takabatake, Y.: The protective role of autophagy against aging and acute ischemic injury in kidney proximal tubular cells. Autophagy **7**(9), 1085–1087 (2011)
57. Jamison, J.M., Gilloteaux, J., Taper, H.S., Calderon, P.B., Summers, J.L.: Autoschizis: a novel cell death. Biochem. Pharmacol. **63**(10), 1773–1783 (2002)
58. Janku, F., McConkey, D.J., Hong, D.S., Kurzrock, R.: Autophagy as a target for anticancer therapy. Nat. Rev. Clin. Oncol. **8**(9), 528–539 (2011)
59. Janssen, A., Medema, R.H.: Entosis: aneuploidy by invasion. Nat. Cell Biol. **13**(3), 199 (2011)
60. Jin, M., Liu, X., Klionsky, D.J., et al.: Snapshot: Selective autophagy. Cell **152**(1–2), 368–368.e2 (2013)
61. Jung, C.H., Ro, S.-H., Cao, J., Otto, N.M., Kim, D.H.: mTOR regulation of autophagy. FEBS Lett. **584**(7), 1287–1295 (2010)
62. Kang, R., Zeh, H.J., Lotze, M.T., Tang, D.: The beclin 1 network regulates autophagy and apoptosis. Cell Death Differ. **18**(4), 571–580 (2011)
63. Kanwar, Y.S., Sun, L.: Shuttling of calcium between endoplasmic reticulum and mitochondria in the renal vasculature. Am. J. Physiol. Ren. Physiol. **295**(5), F1301–F1302 (2008)
64. Karantza-Wadsworth, V., Patel, S., Kravchuk, O., Chen, G., Mathew, R., Jin, S., White, E.: Autophagy mitigates metabolic stress and genome damage in mammary tumorigenesis. Genes Dev. **21**(13), 1621–1635 (2007)
65. Kaushal, G.P., Kaushal, V., Herzog, C., Yang, C.: Autophagy delays apoptosis in renal tubular epithelial cells in cisplatin cytotoxicity. Autophagy **4**(5), 710–712 (2008)
66. Kiffin, R., Bandyopadhyay, U., Cuervo, A.M.: Oxidative stress and autophagy. Antioxid. Redox Signal. **8**(1–2), 152–162 (2006)
67. Kim, I., Rodriguez-Enriquez, S., Lemasters, J.J.: Selective degradation of mitochondria by mitophagy. Arch. Biochem. Biophys. **462**(2), 245–253 (2007)

68. Kim, I., Xu, W., Reed, J.C.: Cell death and endoplasmic reticulum stress: disease relevance and therapeutic opportunities. Nat. Rev. Drug Discov. **7**(12), 1013–1030 (2008)
69. Kim, J., Kundu, M., Viollet, B., Guan, K.-L.: Ampk and mtor regulate autophagy through direct phosphorylation of ulk1. Nat. Cell Biol. **13**(2), 132–141 (2011)
70. King, J.S., Veltman, D.M., Insall, R.H.: The induction of autophagy by mechanical stress. Autophagy **7**(12), 1490–1499 (2011)
71. Klionsky, D.J., Cuervo, A.M., Seglen, P.O., et al.: Methods for monitoring autophagy from yeast to human. Autophagy **3**(3), 181 (2007)
72. Komatsu, M., Waguri, S., Chiba, T., Murata, S., Iwata, J., Tanida, I., Ueno, T., Koike, M., Uchiyama, Y., Kominami, E., et al.: Loss of autophagy in the central nervous system causes neurodegeneration in mice. Nature **441**(7095), 880–884 (2006)
73. Kondo, Y., Kanzawa, T., Sawaya, R., Kondo, S.: The role of autophagy in cancer development and response to therapy. Nat. Rev. Cancer **5**(9), 726–734 (2005)
74. Krick, R., Muehe, Y., Prick, T., Bremer, S., Schlotterhose, P., Eskelinen, E.-L., Millen, J., Goldfarb, D.S., Thumm, M.: Piecemeal microautophagy of the nucleus requires the core macroautophagy genes. Mol. Biol. Cell **19**(10), 4492–4505 (2008)
75. Kristensen, A.R., Schandorff, S., Høyer-Hansen, M., Nielsen, M.O., Jäättelä, M., Dengjel, J., Andersen, J.S.: Ordered organelle degradation during starvation-induced autophagy. Mol. Cell. Proteomics **7**(12), 2419–2428 (2008)
76. Kroemer, G., Levine, B.: Autophagic cell death: the story of a misnomer. Nat. Rev. Mol. Cell Biol. **9**(12), 1004–1010 (2008)
77. Kroemer, G., Mariño, G., Levine, B.: Autophagy and the integrated stress response. Mol. Cell **40**(2), 280–293 (2010)
78. Kumar, V., Abbas, A.K., Fausto, N., Aster, J.C.: Robbins & Cotran Pathologic Basis of Disease, Chap. 7, pp. 25–32. Saunders, Philadelphia (2009)
79. Kumar, V., Abbas, A.K., Fausto, N., Aster, J.C.: Robbins & Cotran Pathologic Basis of Disease, Chap. 7, pp. 303–304. Saunders, Philadelphia (2009)
80. Levine, B., Mizushima, N., Virgin, H.W.: Autophagy in immunity and inflammation. Nature **469**(7330), 323–335 (2011)
81. Li, J.J., Hartono, D., Ong, C.N., Bay, B.H., Yung, L.Y.L.: Autophagy and oxidative stress associated with gold nanoparticles. Biomaterials **31**(23), 5996–6003 (2010)
82. Li, H., Wang, P., Sun, Q., Ding, W.-X., Yin, X.-M., Sobol, R.W., Stolz, D.B., Yu, J., Zhang, L.: Following cytochrome c release, autophagy is inhibited during chemotherapy-induced apoptosis by caspase 8-mediated cleavage of beclin 1. Cancer Res. **71**(10), 3625–3634 (2011)
83. Li, W.-W., Li, J., Bao, J.-K.: Microautophagy: lesser-known self-eating. Cell. Mol. Life Sci. **69**(7), 1125–1136 (2012)
84. Liu, Y., Schiff, M., Czymmek, K., Tallóczy, Z., Levine, B., Dinesh-Kumar, S.P.: Autophagy regulates programmed cell death during the plant innate immune response. Cell **121**(4), 567–577 (2005)
85. Livesey, K.M., Tang, D., Zeh, H.J., Lotze, M.T.: Autophagy inhibition in combination cancer treatment. Curr. Opin. Invest. Drugs **10**(12), 1269–1279 (2009)
86. Luo, S., Rubinsztein, D.C.: Atg5 and bcl-2 provide novel insights into the interplay between apoptosis and autophagy. Cell Death Differ. **14**(7), 1247–1250 (2007)
87. Maiuri, M.C., Zalckvar, E., Kimchi, A., Kroemer, G.: Self-eating and self-killing: crosstalk between autophagy and apoptosis. Nat. Rev. Mol. Cell Biol. **8**(9), 741–752 (2007)
88. Majno, G., Joris, I.: Apoptosis, oncosis, and necrosis. An overview of cell death. Am. J. Pathol. **146**(1), 3 (1995)
89. Mattson, M.P., Chan, S.L.: Calcium orchestrates apoptosis. Nat. Cell Biol. **5**(12), 1041–1043 (2003)
90. Mazure, N.M., Pouysségur, J.: Hypoxia-induced autophagy: cell death or cell survival? Curr. Opin. Cell Biol. **22**(2), 177–180 (2010)
91. Michie, A.M., McCaig, A.M., Nakagawa, R., Vukovic, M.: Death-associated protein kinase (dapk) and signal transduction: regulation in cancer. FEBS J. **277**(1), 74–80 (2010)

92. Mikoshiba, K.: The IP3 receptor/Ca2+ channel and its cellular function. Biochem. Soc. Symp. **74**, 9–22 (2007)
93. Nishida, K., Yamaguchi, O., Otsu, K.: Crosstalk between autophagy and apoptosis in heart disease. Circ. Res. **103**(4), 343–351 (2008)
94. Nixon, R.A., Wegiel, J., Kumar, A., Yu, W.H., Peterhoff, C., Cataldo, A., Cuervo, A.M.: Extensive involvement of autophagy in alzheimer disease: an immuno-electron microscopy study. J. Neuropathol. Exp. Neurol. **64**(2), 113–122 (2005)
95. Notte, A., Leclere, L., Michiels, C.: Autophagy as a mediator of chemotherapy-induced cell death in cancer. Biochem. Pharmacol. **82**(5), 427–434 (2011)
96. Okamoto, K., Kondo-Okamoto, N., Ohsumi, Y.: Mitochondria-anchored receptor ATG32 mediates degradation of mitochondria via selective autophagy. Dev. Cell **17**(1), 87–97 (2009)
97. Orrenius, S., Zhivotovsky, B., Nicotera, P.: Regulation of cell death: the calcium–apoptosis link. Nat. Rev. Mol. Cell Biol. **4**(7), 552–565 (2003)
98. Orvedahl, A., Sumpter, R. Jr., Xiao, G., Ng, A., Zou, Z., Tang, Y., Narimatsu, M., Gilpin, C., Sun, Q., Roth, M., et al.: Image-based genome-wide sirna screen identifies selective autophagy factors. Nature **480**(7375), 113–117 (2011)
99. Ossareh-Nazari, B., Bonizec, M., Cohen, M., Dokudovskaya, S., Delalande, F., Schaeffer, C., Van Dorsselaer, A., Dargemont, C.: CDC48 and UFD3, new partners of the ubiquitin protease UBP3, are required for ribophagy. EMBO Rep. **11**(7), 548–554 (2010)
100. Overholtzer, M., Mailleux, A.A., Mouneimne, G., Normand, G., Schnitt, S.J., King, R.W., Cibas, E.S., Brugge, J.S.: A nonapoptotic cell death process, entosis, that occurs by cell-in-cell invasion. Cell **131**(5), 966–979 (2007)
101. Papandreou, I., Lim, A.L., Laderoute, K., Denko, N.C.: Hypoxia signals autophagy in tumor cells via AMPK activity, independent of HIF-1, BNIP3, and BNIP3l. Cell Death Differ. **15**(10), 1572–1581 (2008)
102. Parmar, J.H., Cook, K.L., Shajahan-Haq, A.N., Clarke, P.A.G., Tavassoly, I., Clarke, R., Tyson, J.J., Baumann, W.T.: Modelling the effect of GRP78 on anti-oestrogen sensitivity and resistance in breast cancer. Interface Focus **3**(4), 20130012 (2013)
103. Pattingre, S., Espert, L., Biard-Piechaczyk, M., Codogno, P.: Regulation of macroautophagy by mtor and beclin 1 complexes. Biochimie **90**(2), 313–323 (2008)
104. Pavlides, S., Whitaker-Menezes, D., Castello-Cros, R., Flomenberg, N., Witkiewicz, A.K., Frank, P.G., Casimiro, M.C., Wang, C., Fortina, P., Addya, S., et al.: The reverse warburg effect: aerobic glycolysis in cancer associated fibroblasts and the tumor stroma. Cell Cycle **8**(23), 3984–4001 (2009)
105. Pavlides, S., Vera, I., Gandara, R., Sneddon, S., Pestell, R.G., Mercier, I., Martinez-Outschoorn, U.E., Whitaker-Menezes, D., Howell, A., Sotgia, F., et al.: Warburg meets autophagy: cancer-associated fibroblasts accelerate tumor growth and metastasis via oxidative stress, mitophagy, and aerobic glycolysis. Antioxid. Redox Signal. **16**(11), 1264–1284 (2012)
106. Periyasamy-Thandavan, S., Jiang, M., Wei, Q., Smith, R., Yin, X.-M., Dong, Z.: Autophagy is cytoprotective during cisplatin injury of renal proximal tubular cells. Kidney Int. **74**(5), 631–640 (2008)
107. Pinton, P., Giorgi, C., Siviero, R., Zecchini, E., Rizzuto, R.: Calcium and apoptosis: Er-mitochondria ca2+ transfer in the control of apoptosis. Oncogene **27**(50), 6407–6418 (2008)
108. Qadir, M.A., Kwok, B., Dragowska, W.H., To, K.H., Le, D., Bally, M.B., Gorski, S.M.: Macroautophagy inhibition sensitizes tamoxifen-resistant breast cancer cells and enhances mitochondrial depolarization. Breast Cancer Res. Treat. **112**(3), 389–403 (2008)
109. Qin, L., Wang, Z., Tao, L., Wang, Y.: Er stress negatively regulates akt/tsc/mtor pathway to enhance autophagy. Autophagy **6**(2), 239–247 (2010)
110. Rabinowitz, J.D., White, E.: Autophagy and metabolism. Sci. Signal. **330**(6009), 1344 (2010)
111. Ravikumar, B., Vacher, C., Berger, Z., Davies, J.E., Luo, S., Oroz, L.G., Scaravilli, F., Easton, D.F., Duden, R., O'Kane, C.J., et al.: Inhibition of mtor induces autophagy and reduces toxicity of polyglutamine expansions in fly and mouse models of huntington disease. Nat. Genet. **36**(6), 585–595 (2004)

112. Reggiori, F., Monastyrska, I., Verheije, M.H., Calì, T., Ulasli, M., Bianchi, S., Bernasconi, R., de Haan, C.A.M., Molinari, M.: Coronaviruses hijack the lc3-i-positive edemosomes, er-derived vesicles exporting short-lived erad regulators, for replication. Cell Host Microbe **7**(6), 500–508 (2010)
113. Rodriguez, D., Rojas-Rivera, D., Hetz, C.: Integrating stress signals at the endoplasmic reticulum: the bcl-2 protein family rheostat. Biochim. Biophys. Acta (BBA)-Mol. Cell Res. **1813**(4), 564–574 (2011)
114. Rong, Y., Distelhorst, C.W.: Bcl-2 protein family members: versatile regulators of calcium signaling in cell survival and apoptosis. Annu. Rev. Physiol. **70**, 73–91 (2008)
115. Seglen, P.O., Gordon, P.B., Holen, I.: Non-selective autophagy. Semin. Cell Biol. **1**, 441–448 (1990)
116. Shang, L., Wang, X.: Ampk and mtor coordinate the regulation of ULK1 and mammalian autophagy initiation. Autophagy **7**(8), 924–926 (2011)
117. Sharma, N., Dey, P.: Cell cannibalism and cancer. Diagn. Cytopathol. **39**(3), 229–233 (2011)
118. Smaili, S.S., Pereira, G.J.S., Costa, M.M., Rocha, K.K., Rodrigues, L., do Carmo, L.G., Hirata, H., Hsu, Y.-T.: The role of calcium stores in apoptosis and autophagy. Curr. Mol. Med. **13**(2), 252–265 (2013)
119. Sotgia, F., Martinez-Outschoorn, U.E., Pavlides, S., Howell, A., Pestell, R.G., Lisanti, M.P., et al.: Understanding the warburg effect and the prognostic value of stromal caveolin-1 as a marker of a lethal tumor microenvironment. Breast Cancer Res. **13**(4), 213 (2011)
120. Sotgia, F., Whitaker-Menezes, D., Martinez-Outschoorn, U.E., Flomenberg, N., Birbe, R., Witkiewicz, A.K., Howell, A., Philp, N.J., Pestell, R.G., Lisanti, M.P.: Mitochondrial metabolism in cancer metastasis: visualizing tumor cell mitochondria and the reverse warburg effect in positive lymph node tissue. Cell Cycle **11**(7), 1445–1454 (2012)
121. Szegezdi, E., MacDonald, D.C., Ní Chonghaile, T., Gupta, S., Samali, A.: BCL-2 family on guard at the ER. Am. J. Physiol. Cell Physiol. **296**(5), C941–C953 (2009)
122. Takahashi, A., Kimura, T., Takabatake, Y., Namba, T., Kaimori, J., Kitamura, H., Matsui, I., Niimura, F., Matsusaka, T., Fujita, N., et al.: Autophagy guards against cisplatin-induced acute kidney injury. Am. J. Pathol. **180**(2), 517–525 (2012)
123. Tallóczy, Z., Virgin H., IV, Levine, B.: PKR-dependent xenophagic degradation of herpes simplex virus type 1. Autophagy **2**(1), 24–29 (2006)
124. Terman, A., Brunk, U.T.: Autophagy in cardiac myocyte homeostasis, aging, and pathology. Cardiovasc. Res. **68**(3), 355–365 (2005)
125. Terman, A., Gustafsson, B., Brunk, U.T.: Autophagy, organelles and ageing. J. Pathol. **211**(2), 134–143 (2007)
126. Twig, G., Elorza, A., Molina, A.J.A., Mohamed, H., Wikstrom, J.D., Walzer, G., Stiles, L., Haigh, S.E., Katz, S., Las, G., et al.: Fission and selective fusion govern mitochondrial segregation and elimination by autophagy. EMBO J. **27**(2), 433–446 (2008)
127. Tyson, J.J., Baumann, W.T., Chen, C., Verdugo, A., Tavassoly, I., Wang, Y., Weiner, L.M., Clarke, R.: Dynamic modelling of oestrogen signalling and cell fate in breast cancer cells. Nat. Rev. Cancer **11**(7), 523–532 (2011)
128. Vandenabeele, P., Galluzzi, L., Berghe, T.V., Kroemer, G.: Molecular mechanisms of necroptosis: an ordered cellular explosion. Nat. Rev. Mol. Cell Biol. **11**(10), 700–714 (2010)
129. Vander Heiden, M.G., Cantley, L.C., Thompson, C.B.: Understanding the warburg effect: the metabolic requirements of cell proliferation. Sci. Signal. **324**(5930), 1029 (2009)
130. Verfaillie, T., Salazar, M., Velasco, G., Agostinis, P.: Linking er stress to autophagy: potential implications for cancer therapy. Int. J. Cell Biol. **2010**, 19 pp. (2010). Article ID 930509
131. Vogt, K.C.: Untersuchungen über die Entwicklungsgeschichte der Geburtshelferkröte (Alytes obstetricans). Jent & Gassmann, Solothurn (1842)
132. Wang, H.-G., Pathan, N., Ethell, I.M., Krajewski, S., Yamaguchi, Y., Shibasaki, F., McKeon, F., Bobo, T., Franke, T.F., Reed, J.C.: Ca2+-induced apoptosis through calcineurin dephosphorylation of bad. Science **284**(5412), 339–343 (1999)

133. Wang, Y., Yu, B., Zhao, J., Guo, J., Li, Y., Han, S., Huang, L., Du, Y., Hong, Y., Tang, D., et al.: Autophagy contributes to leaf starch degradation. Plant Cell Online **25**(4), 1383–1399 (2013)
134. Warburg, O., et al.: On the origin of cancer cells. Science **123**(3191), 309–314 (1956)
135. Wei, Y., Pattingre, S., Sinha, S., Bassik, M., Levine, B.: JNK1-mediated phosphorylation of BCL-2 regulates starvation-induced autophagy. Mol. Cell **30**(6), 678–688 (2008)
136. Wei, Y., Sinha, S.C., Levine, B.: Dual role of jnk1-mediated phosphorylation of bcl-2 in autophagy and apoptosis regulation. Autophagy **4**(7), 949–951 (2008)
137. Weidberg, H., Shvets, E., Elazar, Z.: Lipophagy: selective catabolism designed for lipids. Dev. Cell **16**(5), 628–630 (2009)
138. Weinberg, R.A.: The Biology of Cancer, vol. 1, Chap. 9, p. 353. Garland Science, New York (2007)
139. White, E.: Autophagic cell death unraveled: pharmacological inhibition of apoptosis and autophagy enables necrosis. Autophagy **4**(4), 399–401 (2008)
140. Wise, D.R., Thompson, C.B.: Glutamine addiction: a new therapeutic target in cancer. Trends Biochem. Sci. **35**(8), 427–433 (2010)
141. Xiong, Y., Contento, A.L., Bassham, D.C.: Disruption of autophagy results in constitutive oxidative stress in arabidopsis. Autophagy **3**(3), 257–258 (2007)
142. Xiong, Y., Contento, A.L., Nguyen, P.Q., Bassham, D.C.: Degradation of oxidized proteins by autophagy during oxidative stress in arabidopsis. Plant Physiol. **143**(1), 291–299 (2007)
143. Xu, Y., Yuan, J., Lipinski, M.M.: Live imaging and single-cell analysis reveal differential dynamics of autophagy and apoptosis. Autophagy **9**(9), 1418–1430 (2013)
144. Yamamoto, A., Simonsen, A.: The elimination of accumulated and aggregated proteins: a role for aggrephagy in neurodegeneration. Neurobiol. Dis. **43**(1), 17–28 (2011)
145. Yan, L., Vatner, D.E., Kim, S.-J., Ge, H., Masurekar, M., Massover, W.H., Yang, G., Matsui, Y., Sadoshima, J., Vatner, S.F.: Autophagy in chronically ischemic myocardium. Proc. Natl. Acad. Sci. USA **102**(39), 13807–13812 (2005)
146. Yen, W.-L., Klionsky, D.J.: How to live long and prosper: autophagy, mitochondria, and aging. Physiology **23**(5), 248–262 (2008)
147. Yorimitsu, T., Klionsky, D.J.: Eating the endoplasmic reticulum: quality control by autophagy. Trends Cell Biol. **17**(6), 279–285 (2007)
148. Yousefi, S., Perozzo, R., Schmid, I., Ziemiecki, A., Schaffner, T., Scapozza, L., Brunner, T., Simon, H.U.: Calpain-mediated cleavage of atg5 switches autophagy to apoptosis. Nat. Cell Biol. **8**(10), 1124–1132 (2006)
149. Zalckvar, E., Berissi, H., Eisenstein, M., Kimchi, A.: Phosphorylation of beclin 1 by dap-kinase promotes autophagy by weakening its interactions with BCL-2 and BCL-XL. Autophagy **5**(5), 720–722 (2009)
150. Zalckvar, E., Berissi, H., Mizrachy, L., Idelchuk, Y., Koren, I., Eisenstein, M., Sabanay, H., Pinkas-Kramarski, R., Kimchi, A.: DAP-kinase-mediated phosphorylation on the BH3 domain of beclin 1 promotes dissociation of beclin 1 from BCL-XL and induction of autophagy. EMBO Rep. **10**(3), 285–292 (2009)
151. Zhou, R., Yazdi, A.S., Menu, P., Tschopp, J.: A role for mitochondria in NLRP3 inflammasome activation. Nature **469**(7329), 221–225 (2010)

# Chapter 2
# Mathematical Modeling of the Interplay of Autophagy and Apoptosis

In this chapter, we have proposed a theoretical framework for analysis of dynamics of interplay of autophagy and apoptosis in mammalian cells including cancer cells. Because quantitative experimental data on time course of interplay of autophagy and apoptosis is very limited, we have used the observations of interaction of autophagy and apoptosis in different mammalian cell types to design our primary hypothesis. Using ordinary differential equations (ODEs), we have analyzed network dynamics of molecular signaling pathways controlling cell fate at crosstalk of autophagy and apoptosis. We have used time course of autophagy level and cell fates described by Periyasamy-Thandavan et al. [32] to collectively fit the parameters of the ODE system. The mathematical model presented in this chapter can be extended and by estimating more accurate parameter sets from quantitative experimental data, it can be an integrative *in silico* model of cell fate decision mediated by interplay of autophagy and apoptosis.

## 2.1 Systems Biology of Cell Death Pathways

Extrinsic and intrinsic pathways of apoptosis has been studied at systems-level and mathematical models of cell fate decision mediated by apoptosis have been proposed [2, 5, 34, 35, 41, 42, 44, 56]. Rehm et al. [35] have proposed a mathematical model for spacial and temporal dynamics of MOMP using partial differential equations (PDEs) [35]. ODE-based model of intrinsic pathway of apoptosis have been built by Zhang et al. [56] and Chen et al. [8] proposing apoptosis to be a bistable switch. Extrinsic pathway of apoptosis has been modeled and quantitatively analyzed by Albeck et al. [2, 3, 5].

Taking a systems biology approach to understand the autophagy network and its interplay with other types of cell death has been discussed recently [17, 18, 55]. Martin et al. [24] have presented a computational model of autophagic vesicle

 
I. Tavassoly, *Dynamics of Cell Fate Decision Mediated by the Interplay of Autophagy and Apoptosis in Cancer Cells*, Springer Theses, DOI 10.1007/978-3-319-14962-2_2

dynamics in single cells, but they did not address the interplay of autophagy and apoptosis. Kapuy et al. [19] have addressed the crosstalk between autophagy and apoptosis using a simple mathematical model, but they have not compared their simulations with experimental measurements of how live cells respond to stress. In their model they have used a very simple protein interaction network which does not consider some important components linking autophagy to apoptosis. Their model has not considered BCL2 proteins in different compartments such as ER and mitochondria [19]. In this model there is no link through any signaling pathway to connect stress to basic components controlling autophagy and apoptosis. One example can be role of DAPK initiating autophagy without any effect on BCL2 proteins in ER, which has been missed in the influence diagram used for this model. In the model proposed by Kapuy et al. [19], autophagy as a process has not been taken into account and the network structure of the very basic signaling pathways controlling autophagy and apoptosis is responsible for giving a delay for initiation of bistable switch of apoptosis. In other words in this model autophagy is just defined as existence of one protein component, Beclin-1, in the signaling network, and the role of autophagy as a dynamic cellular process to inhibit the stress is missed.

High-throughput systems biology has revealed global protein interaction network controlling autophagy in mammalian cells and has provided a global interaction landscape of autophagy network [7]. Recently Xu et al. [50] have provided an experimental methodology for temporal analysis of interplay of autophagy and apoptosis in live cells.

## 2.2 Dynamic Modeling of the Interplay of Autophagy and Apoptosis

The basic hypothesis based on which the dynamic model of autophagy and apoptosis was built, had come from previous basic observations in different cell lines and model systems. These observations are summarized in Fig. 2.1 (this figure was made by online protein lounge tools).

Most mammalian cells, including cancer cells present a level of constitutive (basal) autophagy under physiological condition [20, 21, 28, 29, 39]. In some cases when a cell faces stress, autophagy flux is increased to keep the survival and avoid cell death [23, 30].

Another scenario is when a cell increases its basal level of autophagy, when it is dealing with a stress, but finally it commits cell death. In this scenario, increased autophagy level causes cell death to happen later. Knockdown of autophagy machinery will cause cell death to happen sooner [1, 6, 23]. Blocking both autophagy machinery and apoptosis pathway under stress conditions ends in necrosis [47] (Fig. 2.2).

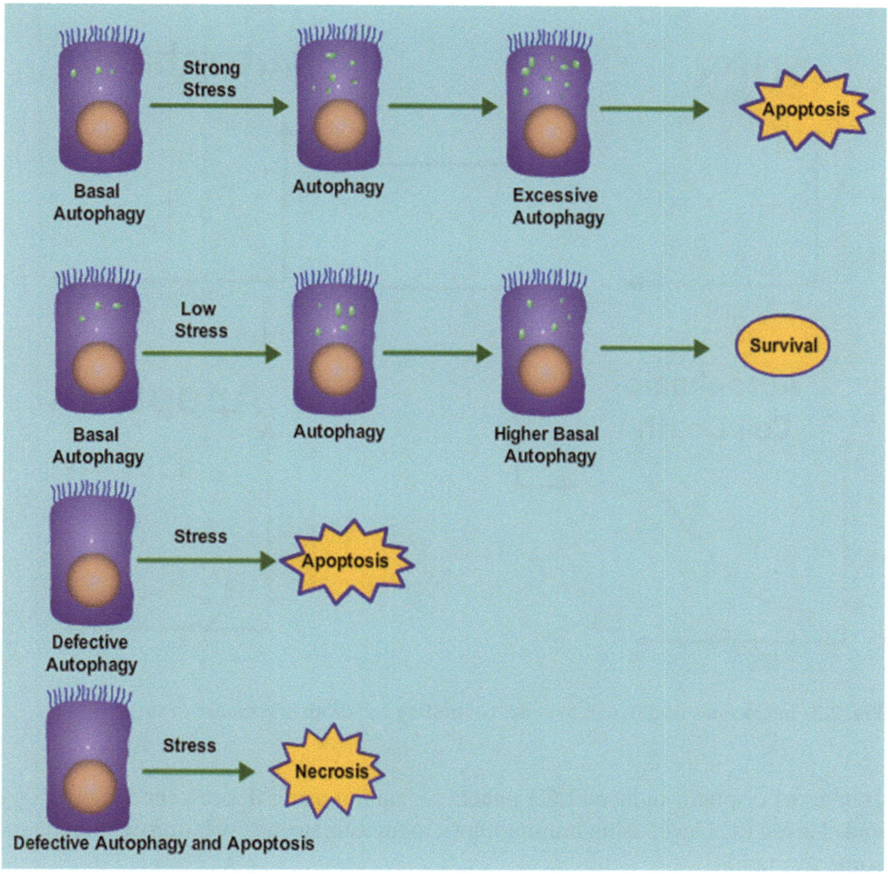


**Fig. 2.1** Different cell fate scenarios mediated by autophagy

## *2.2.1 Mathematical Formalism*

To integrate the knowledge contained in the literature and determine if it can account for the qualitative and quantitative results of experiments in a consistent manner, we built a mathematical model of the molecular mechanism controlling autophagy and apoptosis. The experimental data we seek to explain involves the response of cells, in terms of autophagy and apoptosis, to treatment with cisplatin, a cytotoxic drug. Cisplatin induces UPR and activation of JNK and DAPK in mammalian cells [11, 38, 49].

Periyasamy-Thandavan et al. [32] used rat kidney proximal tubule cells (RPTCs), transiently transfected with GFP-LC3 and treated with cisplatin, to investigate the cytoprotective role of autophagy. As part of their study, they measured time

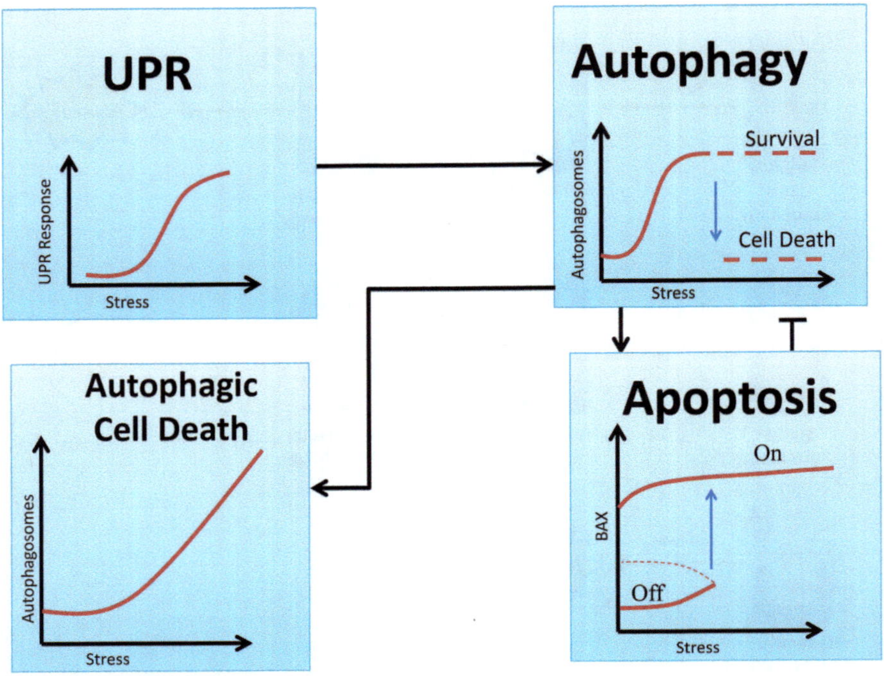


**Fig. 2.2** Interactions of different modules controlling cell death in response to stress

courses of cisplatin-induced LC3 puncta formation using fluorescence microscopy and cleaved LC3 level using immunoblots, to investigate the evolution of autophagosome formation.

As in any modeling task, we make certain simplifying assumptions to reduce the complexity of the model while capturing the principal molecular details related to the experimental data under consideration. To this end, we have based our mathematical model on the influence diagram in Fig. 2.3.

In this diagram, simple modules for autophagy and apoptosis are connected to summarize the main components of stress regulation in mammalian cells. Our apoptosis module is inspired by a mathematical model of the intrinsic pathway proposed by Zhang et al. [56], who showed that stoichiometric binding between anti-apoptotic BCL2 proteins and pro-apoptotic BH3 proteins creates an on/off (bistable) switch for MOMP and subsequent activation of apoptotic proteases. As soon as BH3 level exceeds BCL2 level, the switch turns on. We assume that stress up-regulates BH3 production by a pathway dependent on autophagy via release of calcium into the cytoplasm, and a second pathway independent of autophagy. While the mechanisms by which external stress (such as cisplatin) activates autophagy and apoptosis are unclear, we assume that stress is communicated to these subsystems by activating JNK and DAPK. In turn, JNK phosphorylates BCL2 family proteins in the ER [30, 45, 46] and DAPK phosphorylates Beclin-1 [26, 51, 53, 54].

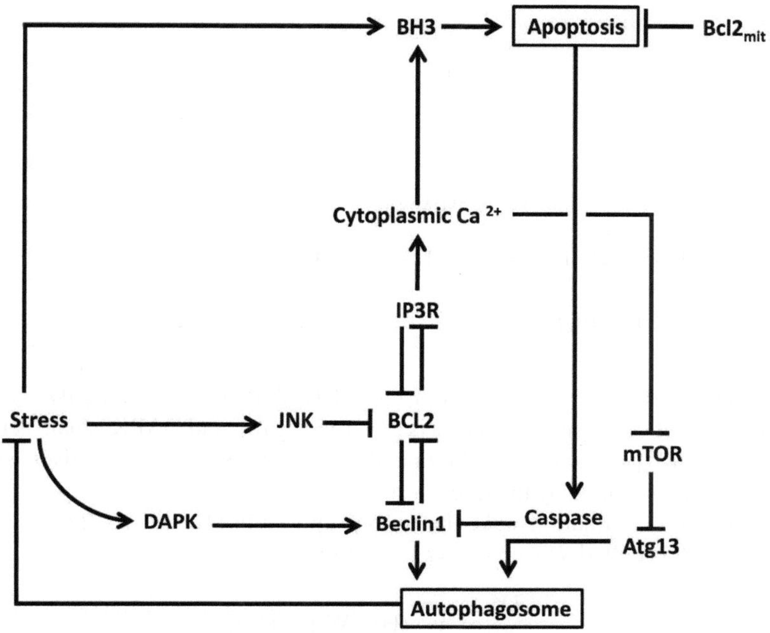


**Fig. 2.3** An influence diagram of the molecular interactions controlling autophagosome formation and the interplay between autophagy and apoptosis

We model the internal stress of the cell with the equation

$$\frac{dS}{dt} = k_c \cdot C \cdot \mathrm{Heav}(t) + k_{\mathrm{sb}} - (k_{\mathrm{rb}} + k_{\mathrm{ra}}.[\mathrm{Atphg}]) \cdot S$$

Cisplatin treatment causes internal stress to increase at a rate proportional to cisplatin concentration, $C$ and stress is relieved by autophagy (rate constant, $k_{\mathrm{ra}}$). We also assume basal rates of production ($k_{\mathrm{sb}}$) and relief ($k_{\mathrm{rb}}$) of stress, in order to set a basal level of autophagy. By defining an ODE for stress, we can mathematically define severe stress (increasing cisplatin dose) and prolonged stress (increasing the time of cisplatin treatment). Because the molecular mechanisms of many of the processes we seek to model are unclear or unknown, we have adopted a phenomenological approach to modeling the interaction diagram in Fig. 2.3. For example, to model the activation of DAPK we use the equation

$$\frac{d[\mathrm{DAPK}]}{dt} = \gamma_{\mathrm{D}} \cdot (\mathrm{H}(\sigma_{\mathrm{D}} \mathrm{W}_{\mathrm{D}}) - [\mathrm{DAPK}])$$

where

$$H(\sigma_D W_D) = \frac{1}{1 + e^{-\sigma_D W_D}}$$

and

$$W_D = w_{DAPK\_0} + w_{DAPK\_S} \cdot S$$

The steady-state level of active DAPK is given by the sigmoidal function

$$[DAPK]_{ss} = \frac{1}{1 + e^{-\sigma_D} \cdot (w_{DAPK\_0} + w_{DAPK\_S} \cdot S)}$$

which varies between 0 (no activation) and 1 (full activation). Roughly speaking, the steepness of the sigmoid function is governed by $\sigma_D$, and DAPK is half-activated when $S = -w_{DAPK\_0}/w_{DAPK\_S}$. The rate at which DAPK moves towards its steady-state value is controlled by the rate constant $\gamma_D$.

In general, our equations take the form proposed in [10, 12, 16, 27, 33, 40, 43, 44, 48] which is called Wilson–Cowan equation

$$\frac{dX_i}{dt} = \gamma_i \cdot (H(\sigma_i W_i) - [X_i])$$

$$W_i = w_{i0} + \sum_j w_{ij} X_j$$

where the subscript i refers to the species in our model (DAPK, JNK, mTOR, etc.), and $X_i$ is the concentration of the active form of species i. $\gamma_i$ is a rate constant determining how fast $X_i$ approaches its steady-state value . The function $H(\sigma_i W_i)$ is sigmoidally shaped, varying from 0 for $W_i << 1/\sigma_i$ to 1 for $W_i >> 1/\sigma_i$. The parameter $\sigma_i$ determines the steepness of the sigmoid as a function of W near $W = 0$. $W_i$ is the net effect on protein i from all other proteins interacting with it; $w_{ij} < 0$ for inhibitory interactions, $w_{ij} > 0$ for activating interactions, and $w_{i0}$ determines the position of the switch when all protein activities are 0.

One of the advantages of Wilson–Cowan equation is its ability to let model interactions among several components, when the details and exact mechanism of the interactions are not clear. An example is the activation of JNK and DAPK by stress (cisplatin). We know that upon cisplatin treatment, JNK and DAPK are activated but the exact molecular mechanism linking cytotoxic effects of cisplatin like DNA damage to DAPK and JNK activation is not clear.

There are a few exceptions to this formalism in our model; namely, the ODEs for the cleavage of Beclin-1, for the change in autophagosome concentration, and for the movement of calcium between the ER and the cytoplasm. We write these rate equations as elementary first- and second-order chemical reactions. The nonlinear ODEs and related algebraic equations we use to model the influence diagram in Fig. 2.3 are presented in Tables 2.1 and 2.2, and the variables and parameter values are shown in Tables 2.3, and 2.4 respectively.

**Table 2.1** Ordinary differential equations

| |
|---|
| $\frac{dS}{dt} = k_c \cdot C \cdot \mathrm{Heav}(t) + k_{\mathrm{sb}} - (k_{\mathrm{rb}} + k_{\mathrm{ra}}.[\mathrm{Atphg}]) \cdot S$ |
| $\frac{d[\mathrm{Atphg}]}{dt} = k_a \cdot ([\mathrm{BECN1}]_\mathrm{F} + [\mathrm{BECN1\_P}])[\mathrm{ATG13}] - k_{\mathrm{da}} \cdot [\mathrm{Atphg}]$ |
| $\frac{d[\mathrm{JNK}]}{dt} = \gamma_\mathrm{J} \cdot (\mathrm{H}(\sigma_\mathrm{J} \mathrm{W_J}) - [\mathrm{JNK}])$ |
| $\frac{d[\mathrm{DAPK}]}{dt} = \gamma_\mathrm{D} \cdot (\mathrm{H}(\sigma_\mathrm{D} \mathrm{W_D}) - [\mathrm{DAPK}])$ |
| $\frac{d[\mathrm{BCL2\_P}]}{dt} = \gamma_\mathrm{B} \cdot ([\mathrm{BCL2}]_\mathrm{T} \cdot \mathrm{H}(\sigma_\mathrm{B} \mathrm{W_B}) - [\mathrm{BCL2\_P}])$ |
| $\frac{d[\mathrm{Ca}^{2+}]}{dt} = k_{\mathrm{out}} \cdot ([\mathrm{Ca}^{2+}]_\mathrm{T} - [\mathrm{Ca}^{2+}]) - k_{\mathrm{in}} \cdot [\mathrm{Ca}^{2+}]$ |
| $\frac{d[\mathrm{mTOR}]}{dt} = \gamma_\mathrm{T} \cdot (\mathrm{H}(\sigma_\mathrm{T} \mathrm{W_T}) - [\mathrm{mTOR}])$ |
| $\frac{d[\mathrm{ATG13}]}{dt} = \gamma_\mathrm{G} \cdot (\mathrm{H}(\sigma_\mathrm{G} \mathrm{W_G}) - [\mathrm{ATG13}])$ |
| $\frac{d[\mathrm{BH3}]}{dt} = \gamma_\mathrm{H} \cdot (\mathrm{H}(\sigma_\mathrm{H} \mathrm{W_H}) - [\mathrm{BH3}])$ |
| $\frac{d[\mathrm{CASP}]}{dt} = \gamma_\mathrm{C} \cdot (\mathrm{Heav}([\mathrm{BH3}] - [\mathrm{BCL2}]_{\mathrm{mit}}) - [\mathrm{CASP}])$ |
| $\frac{d[\mathrm{BECN1}]_\mathrm{T}}{dt} = -k_{\mathrm{CASP}} \cdot [\mathrm{CASP}] \cdot [\mathrm{BECN1}]_\mathrm{T}$ |
| $\frac{d[\mathrm{BECN1\_P}]}{dt} = \gamma_\mathrm{L} \cdot ([\mathrm{BECN1}]_\mathrm{u} \cdot \mathrm{H}(\sigma_\mathrm{L} \mathrm{W_L}) - [\mathrm{BECN1\_P}])$ |

**Table 2.2** Algebraic equations

| |
|---|
| $\mathrm{W_J} = w_{\mathrm{JNK\_0}} + w_{\mathrm{JNK\_S}} \cdot S$ |
| $\mathrm{W_D} = w_{\mathrm{DAPK\_0}} + w_{\mathrm{DAPK\_S}} \cdot S$ |
| $\mathrm{W_B} = w_{\mathrm{BCL2\_0}} + w_{\mathrm{BCL2\_JNK}} \cdot [\mathrm{JNK}]$ |
| $\mathrm{W_T} = w_{\mathrm{mTOR\_0}} + w_{\mathrm{mTOR\_Ca}} \cdot [\mathrm{Ca}^{2+}]$ |
| $\mathrm{W_G} = w_{\mathrm{ATG\_0}} + w_{\mathrm{ATG\_mTOR}} \cdot [\mathrm{mTOR}]$ |
| $\mathrm{W_H} = w_{\mathrm{BH3\_0}} + w_{\mathrm{BH3\_Ca}} \cdot [\mathrm{Ca}^{2+}] + w_{\mathrm{BH3\_S}} \cdot S$ |
| $\mathrm{W_L} = w_{\mathrm{BECN1\_0}} + w_{\mathrm{BECN1\_DAPK}} \cdot [\mathrm{DAPK}]$ |
| $[\mathrm{BCL2}]_\mathrm{u} = [\mathrm{BCL2}]_\mathrm{T} - [\mathrm{BCL2\_P}]$ |
| $[\mathrm{BECN1}]_\mathrm{u} = [\mathrm{BECN1}]_\mathrm{T} - [\mathrm{BECN1\_P}]$ |
| $[\mathrm{LIG}]_\mathrm{T} = [\mathrm{IP3R}]_\mathrm{T} + [\mathrm{BECN1}]_\mathrm{u}$ |
| $[\mathrm{LIG}]_\mathrm{F} = \max(0, [\mathrm{LIG}]_\mathrm{T} - [\mathrm{BCL2}]_\mathrm{u})$ |
| $[\mathrm{IP3R}]_\mathrm{F} = [\mathrm{IP3R}]_\mathrm{T} \cdot \frac{[\mathrm{LIG}]_\mathrm{F}}{[\mathrm{LIG}]_\mathrm{T}}$ |
| $[\mathrm{BECN1}]_\mathrm{F} = [\mathrm{BECN1}]_\mathrm{u} \cdot \frac{[\mathrm{LIG}]_\mathrm{F}}{[\mathrm{LIG}]_\mathrm{T}}$ |

The equations in Table 2.1 can be classified as those dealing with stress induction, autophagosome formation, and apoptosis initiation. Autophagosome formation is controlled by the levels of active ATG13 and Beclin-1, and the removal of autophagosomes refers to their docking to lysosomes.

At the ER membrane, the interactions among BCL2 proteins, Beclin-1 and IP3R contribute to the complex interplay of autophagy and apoptosis. DAPK phosphorylates Beclin-1, causing dissociation from BCL2 and initiation of autophagosome formation. Newly available BCL2 can bind to IP3R, reducing calcium release from the ER and postponing apoptosis. Active JNK phosphorylates BCL2, inhibiting its ability to bind to IP3R or Beclin-1, and hence promoting both apoptosis and autophagy. In addition to the ODEs, there are several algebraic equations

**Table 2.3** Variables, their descriptions, and values when there is no stress

| Variable | Description | Steady-state value (no drug treatment) |
|---|---|---|
| $S$ | Level of stress induced in the cell by drug treatment or other stressors | 0.03 |
| [Atphg] | Concentration of autophagosomes in cytoplasm | 0.774 |
| [JNK] | Concentration of active c-Jun N-terminal kinase | 0.079 |
| [DAPK] | Concentration of active death-associated protein kinase | 0.046 |
| [BCL2_P] | Concentration of phosphorylated BCL2 family proteins in ER | 0.573 |
| $[BCL2]_u$ | Concentration of un-phosphorylated BCL2 family proteins in ER | 2.427 |
| $[LIG]_T$ | Concentration of total ligands available for binding to BCL2 family proteins in ER | 3.102 |
| $[LIG]_F$ | Concentration of ligands free from suppression by BCL2 family proteins | 0.675 |
| $[BECN1]_T$ | Concentration of total Beclin-1 protein | 3 |
| [BECN1_P] | Concentration of phosphorylated Beclin-1 protein | 0.898 |
| $[BECN1]_u$ | Concentration of un-phosphorylated form of Beclin-1 protein | 2.102 |
| $[BECN1]_F$ | Concentration of Beclin-1 protein which is free from suppression by BCL2 | 0.457 |
| $[Ca^{2+}]$ | Concentration of cytoplasmic $Ca^{2+}$ | 0.196 |
| [mTOR] | Concentration of active mammalian target of rapamycin (mTOR) protein | 0.592 |
| [ATG13] | Concentration of active ATG13 protein | 0.085 |
| [BH3] | Concentration of active BH3 proteins | 0.055 |
| [CASP] | Concentration of active caspase | 0 |

representing association and dissociation reactions. The binding of BCL2 to IP3R and to Beclin-1 in the ER is assumed to be so strong that any free BCL2 will immediately bind to any unbound IP3R or free (unphosphorylated and uncleaved) Beclin-1. If BCL2 is not in excess of its potential binding partners, the amounts of unbound IP3R and free Beclin-1 are allocated proportionally. The algebraic equations representing the fast binding reactions are given in Table 2.2. In these equations, $[LIG]_T$ represents total concentration of proteins, either IP3R or Beclin-1, which have the potential to bind to BCL2 family proteins in the ER.

Apoptosis is represented by the variable [CASP] $=$ concentration of active caspase. [CASP] is governed by a piecewise linear ODE with steady-state $= 0$ when $[BCL2]_{mit} \geq [BH3]$ and steady-state $= 1$ when $[BH3] \geq [BCL2]_{mit}$ . BH3 production is driven principally by cytoplasmic calcium. Hence, if $[Ca^{2+}]$ is elevated a sufficiently long time, enough BH3 will accumulate to flip the caspase switch from 0 to 1. Active caspase then cleaves Beclin-1 and inhibits further autophagosome formation.

**Table 2.4** Parameters, their descriptions, and their numerical values

| Parameters | Description | Value |
|---|---|---|
| $k_a, k_{da}$ | Rate constants for autophagosome formation and degradation ($h^{-1}$) | 2,0.3 |
| $C$ | Drug (cisplatin) dose ($\mu$M) | variable |
| $k_{sb}$ | Basal rate of stress ($h^{-1}$) | 0.1 |
| $k_c$ | Drug-induced stress rate ($\mu M^{-1} h^{-1}$) | 0.25 |
| $k_{rb}$ | Rate constant for background relief of stress ($h^{-1}$) | 1 |
| $k_{ra}$ | Rate constant for autophagic relief of stress ($h^{-1}$) | $10\,k_{da}$ |
| $\gamma_J, \gamma_D, \gamma_B, \gamma_T$ | Rate constants for changes in protein concentration ($h^{-1}$) | 1,1,0.5,1 |
| $\gamma_G, \gamma_H, \gamma_C, \gamma_L$ | | 1,0.2,1,10 |
| $\sigma_J, \sigma_D, \sigma_B, \sigma_H$ | Steepness of sigmoidal response curves | 5,10,2,1 |
| $\sigma_G, \sigma_T, \sigma_L$ | | 4,10,1 |
| $k_{CASP}$ | Rate constant for cleavage of Beclin-1 by caspase ($h^{-1}$) | 2.5 |
| $k_{out}, k_{in}$ | Rate constant for $Ca^{2+}$ transport from ER to cytoplasm and vice versa ($h^{-1}$) | 0.5,1 |
| $w_{JNK\_0}, w_{DAPK\_0}$ | Offset of sigmoidal function when there are no signals | −0.5,−0.3 |
| $w_{BCL2\_0}, w_{mTOR\_0}$ | | −0.8,2 |
| $w_{ATG13\_0}, w_{BH3\_0}$ | | 0,−5 |
| $w_{BECN1\_0}$ | | −0.3 |
| $w_{JNK\_S}, w_{DAPK\_S}$ | Interaction coefficient | 0.3,0.1 |
| $w_{mTOR\_Ca}, w_{ATG13\_mTOR}$ | | −10,−1 |
| $w_{BH3\_Ca}, w_{BH3\_S}$ | | 11,0.1 |
| $w_{BECN1\_DAPK}, w_{BCL2\_JNK}$ | | 0.2,1 |
| $[BCL2]_T$ | Total BCL2 family proteins in ER | 3 |
| $[IP3R]_T$ | Total IP3R proteins in ER | 1 |
| $[BCL2]_{mit}$ | Total anti-apoptotic BCL2 family proteins in mitochondria | Randomly selected from a normal distribution with mean = 0.1 and SD = 0.03 |
| $[Ca^{2+}]_T$ | Maximum cytoplasmic $[Ca^{2+}]$ due to release of ER calcium | 2 |

In our simulations, we provide a basal level of autophagy by assuming the cell has a basal level of stress under physiological conditions.

Tumors and cancer cell lines are heterogeneous cell populations. Some of this heterogeneity originates from fluctuations in cellular components and from differences in cellular states such as cell cycle phases. The tumors microenvironment also has a significant effect on this heterogeneity. Experimental methods such as western blotting, however, measure the average behavior of populations of cells [15, 22, 25, 52]. To introduce cell heterogeneity and population averaging into our model, we assume a normal distribution (mean = 0.1 and standard deviation = 0.03) of BCL2 anti-apoptosis proteins in mitochondria, and we randomly simulate 100 cells with different mitochondrial BCL2 levels. This approach allows us to calculate the percentage of cells undergoing apoptosis and the time course of average LC3-II level (as an autophagy marker). Spencer et al. [42] have confirmed that fluctuations in protein components controlling apoptosis play important roles in the probability and timing of programmed cell death.

### *2.2.2 Results*

Little quantitative time course data pertaining to autophagy is currently available in the literature. To test our proposed model structure and tune our parameter values, we use measurements by Periyasamy-Thandavan et al. [32] of autophagy and apoptosis in RPTCs in response to cisplatin treatment, which induces the UPR and activates JNK and DAPK in mammalian cells. Cisplatin treatment also inhibits mTOR and induces autophagy as a protective mechanism in cancer cells [11, 14, 38, 49].

Cisplatin is one of the most common cytotoxic chemotherapy agents which is widely used in oncology practice. The main mechanism of action of cisplatin is through DNA damage [31, 37]. DNA damage is one of the inducers of autophagy and autophagy is a pro-survival response after DNA damage [1, 36]. But the exact molecular mechanisms linking DNA damage to autophagy is poorly understood [36]. Crighton et al. [9] have found that P53 activates autophagy via DRAM (damage-regulated autophagy modulator) upon DNA damage.

Cisplatin treatment triggers autophagic response in glioma cells as a dose- and time-dependent protective response [14]. Real-time tracking of cellular and molecular alterations in HT-29, HCT-116, HepG2, and MCF-7 cells has shown that apoptosis is triggered in these cell lines about 8–11 h after cisplatin treatment [4]. After 5–6 h of 50 μM cisplatin treatment MCF-7 cells go through diminished mitochondrial respiration. Then they show an increased level of glycolysis about 8–9 h after cisplatin treatment. Finally cell death starts in them after 10–11 h [4].

Treating RPTCs with 20 μM cisplatin, Periyasamy-Thandavan et al. [32] measured cisplatin-induced autophagy at five time points over 24 h using immunoblot analysis of LC3-II. They found that the LC3-II level rose four-fold and then decayed to the original level.

### Autophagy as a Rheostat

The first response to cisplatin-induced stress is to activate autophagy, in an attempt to relieve the stress. If stress (cisplatin dose) is too large, autophagy cannot rescue the cell, and it proceeds towards cell death. Figure 2.4a shows simulated time courses of autophagy as predicted by the model. Increasing doses of cisplatin have been applied (0, 4, 8, 12, 16, 20, 24, 28 μM). When there is no cisplatin treatment, autophagy stays at its basal level. The gradual increase in steady-state level of autophagy with increasing dose of cisplatin, we call the rheostat function of autophagy, which is pro-survival and anti-apoptotic. For large doses of cisplatin the steady-state level of autophagy eventually drops to zero because the final cell fate in these cases is apoptosis, and the cleavage of Beclin-1 by caspase 8 shuts off autophagy. The stress–response curve (Fig. 2.4b) plots the final steady-state level of autophagy as a function of cisplatin dose. The rheostat function of autophagy is evident for low stress levels, but when stress reaches the point-of-no-return, apoptosis switches on. Even in the latter case, autophagy works to suppress the stress and to postpone the initiation of cell death. In other words, the role of autophagy is to block cell death and promote survival, but when stress is too strong, apoptosis will eventually be triggered and autophagy will be shut down. Figure 2.5a shows the time course of autophagy in RPTCs treated with 20 μM cisplatin over the course of 24 h. Initially, autophagy increases, but then the level reduces as the cells commit apoptosis. Our model simulation (solid line) is in reasonable agreement with the experimental data (green squares). The model simulation averages over 100 simulations of cells with different levels of mitochondrial BCL2 anti-apoptotic proteins. We use this averaging procedure because the experimental data points average over the activities in many cells and because of the intrinsic heterogeneity of individual mammalian cells. This approach allows us to predict the percentage of apoptosis in RPTCs (Table 2.5). Interestingly after 12 h treatment with 20 μM cisplatin, some cells have committed apoptosis and some cells have survived, which is due to the pro-survival role of autophagy and to the different levels of mitochondrial anti-apoptotic proteins in different cells.

### Overexpression of BCL2 and Cell Fate

BCL2 proteins in the ER and mitochondria regulate the interplay between autophagy and apoptosis. Overexpression of BCL2 proteins blocks autophagy because Beclin-1 cannot be released from BCL2 inhibition. At the same time, cell death is inhibited because BCL2 in mitochondria blocks the activation of BAX, the trigger of apoptosis. Figure 2.5b shows a time course of autophagy in RPTCs overexpressing BCL2 and treated with 20 μM cisplatin. In both experiments and computations, the basal level of autophagy is decreased and autophagy is suppressed. For the computational simulation, the parameters controlling BCL2 proteins in the ER and mitochondria have been increased two-fold.

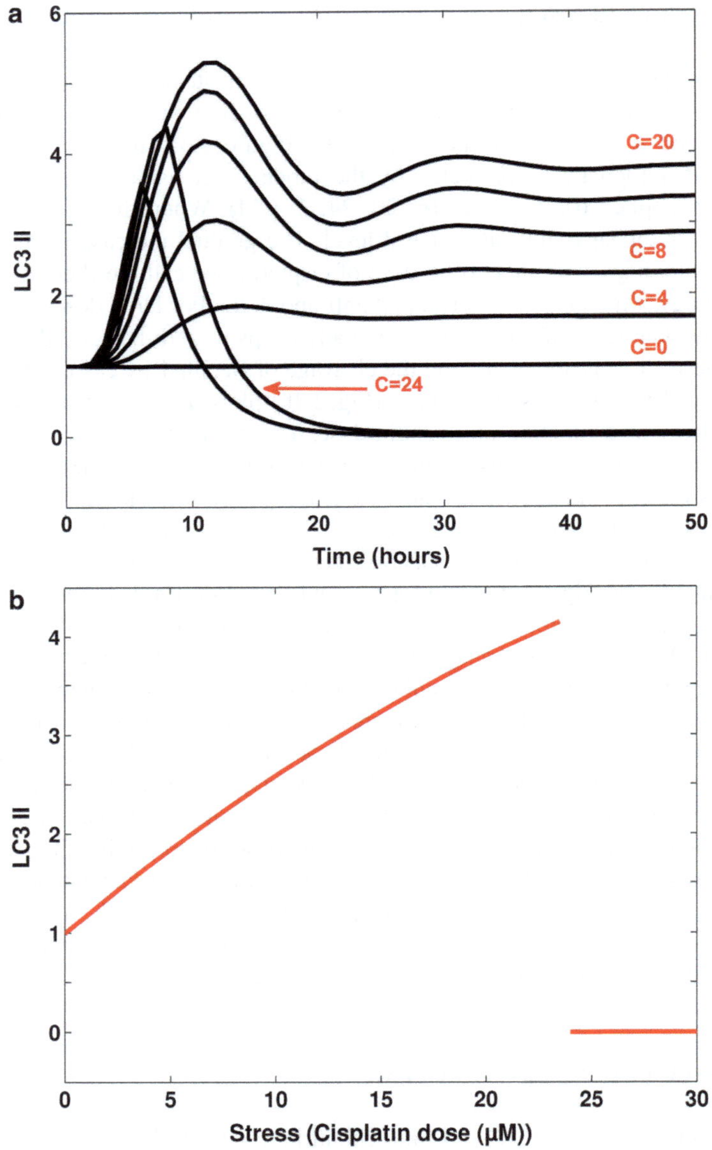


**Fig. 2.4** Rheostat function of autophagy. (**a**) Time-course of LC3 II for different doses of cisplatin. (**b**) Dose/Response (Autophagy) Curve

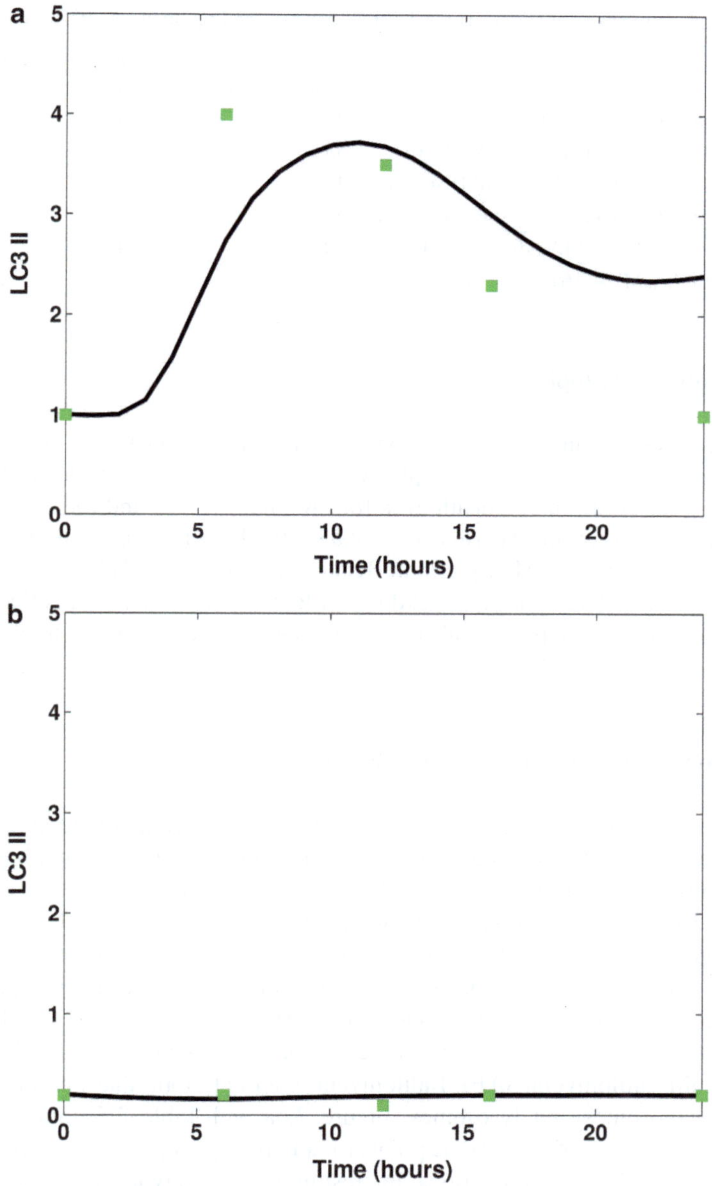


**Fig. 2.5** Time course of autophagy level under cisplatin treatment without (panel **a**) with (panel **b**) BCL2 overexpression

**Table 2.5** Percentage of apoptosis for different experimental regimens *in vitro* and *in silico*

| Conditions | Experimental measurement (%) | Model prediction (%) |
|---|---|---|
| Cisplatin (20 μM) @ 12 h | 27 | 33 |
| Cisplatin (20 μM)+BAF (100 nM) @ 12 h | 55 | 53 |
| Cisplatin (20 μM)+3-MA (10 mM) @ 12 h | 47 | 51 |
| Cisplatin (20 μM)+Beclin-1 shRNA @ 16 h | 52 | 50 |
| Cisplatin (5 μM)+3-MA (10 mM) @ 16 h | 20 | 14 |
| Cisplatin (5 μM) @ 16 h | 3 | 10 |

### Knockdown of Autophagy

Under any conditions, the knockdown of autophagy machinery will lead to increased cell death, because autophagy is pro-survival and anti-death. Crippled autophagy can result in cell death even for low stress levels, and for high levels of stress, apoptosis is initiated earlier. Table 2.5 shows the percentage of apoptosis after 16 h treatment with 20 μM cisplatin in Beclin-1 knockdown RPTCs. As Beclin-1 is a main regulator of autophagy, knockdown of Beclin-1 has a dominant effect on cell death rate. For the computational simulation, the initial condition of total Beclin-1 was reduced from 3 to 2.

### Inhibition of Autophagosome Degradation

The key role of autophagy in cell survival comes from docking of autophagosomes with lysosomes and formation of autolysosomes. Degradation of autophagosomes in this process will provide cells with the raw materials and ATP they need to survive. Blocking the formation of autolysosomes will increase the number of autophagosomes in the cell. At the same time cell death is increased because autophagy cannot serve its pro-survival function. By decreasing the parameter controlling autophagosome degradation from 0.3 to 0.2, we have simulated the average time course of autophagy in cells that are treated with 20 μM cisplatin and 100 nM bafilomycin(BAF). Bafilomycin is an antibiotic that inhibits docking of autophagosomes and lysosomes. Figure 2.6a and Table 2.5 compare model predictions with experimental data points for autophagy progression and percentage of apoptosis under these conditions. Interestingly autophagy level stays high while cell death rate increases.

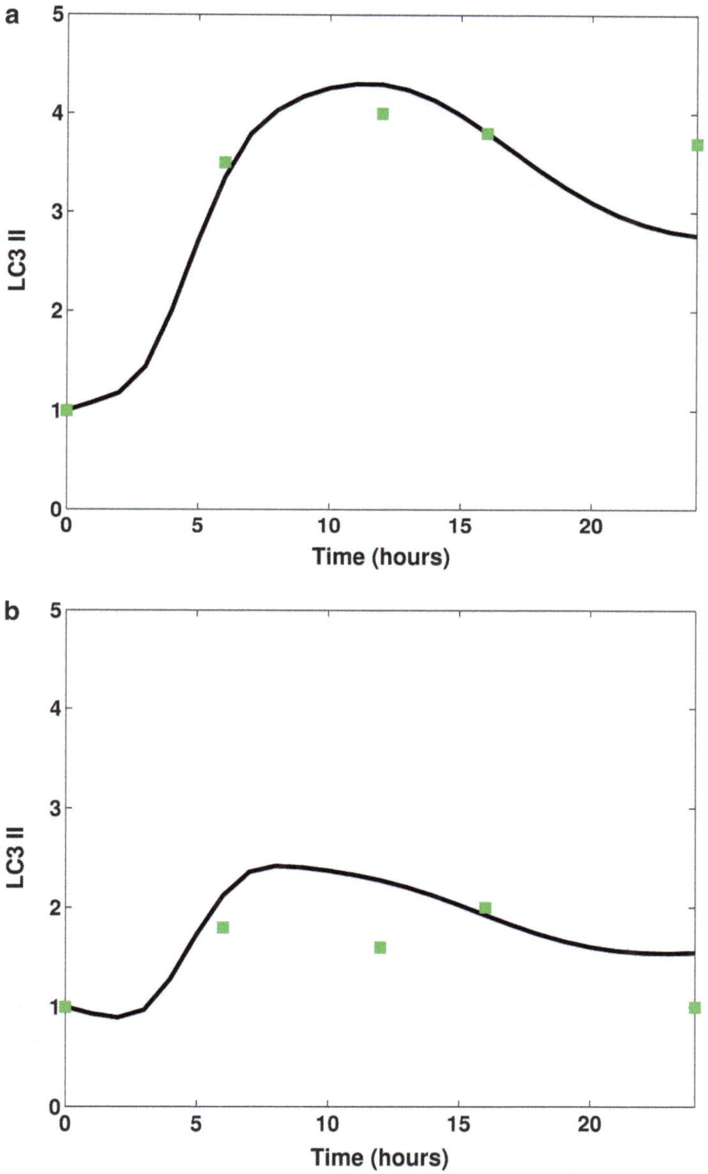


**Fig. 2.6** Time course of autophagy level under cisplatin treatment combined with BAF (panel **a**) and 3-MA (panel **b**)

**Inhibition of Autophagosome Formation**

3-Methyladenine (3-MA) is a drug that inhibits formation of autophagosomes. Intuitively, treating cells with 3-MA should decrease autophagy and increase cell death. Experimentally RPTCs respond to 10 mM 3-MA and 20 μM cisplatin with a higher percentage of apoptosis and a decreased level of autophagy; our model shows the same behavior (Table 2.5 and Fig. 2.6b). In this case, the parameter controlling formation of autophagosomes has been decreased from 2 to 1.5. Table 2.4 also shows predictions of the model for 16 h treatment of RPTCs with 5 μM cisplatin alone and in combination with 3-MA. Adding 3-MA significantly increases the cell death rate because it stops autophagy at the very first steps.

## 2.2.3 *Future Directions*

Assigning the accurate parameters in mathematical models of cellular control systems can be done using direct experimental measurements. But quantitative experimental measurements are very hard to do especially when dealing with systems of ODEs with high number of parameters. Collective fitting of parameters to experimental data (what we have done in present mathematical model) is a good way to assign parameters. But in systems biology, all models show sloppiness of parameter sensitivities. In other words there are always more than one parameter set which can be assigned to model to capture the dynamics of experimental time-series data [13]. It means that for each model, there is a parameter space within which the model can reproduce the experimental time-series data. One next step for current modeling framework is to evaluate the sloppiness of the parameters and calculate the parameter space for which dynamics showed by the model is robust.

Another step is to add other important molecular components to the wiring diagram of interplay of autophagy and apoptosis. This molecular components include P53, molecular signaling pathway of UPR, control system of programmed necrosis, and cell cycle signaling network. Tyson et al. [44] have proposed a roadmap to connect modules of apoptosis, UPR, autophagy, cell cycle, and other signaling pathways in breast cancer to build a virtual *in silico* model of cell signaling network in cancer cells.

Experimental measurements of autophagy and apoptosis temporal behaviors need to be done in single cells and cell populations and even in tumor to validate the model and find the optimized parameter space for different types of cells. We discuss these experimental methodologies in Chap. 3.

# References


1. Abedin, M.J., Wang, D., McDonnell, M.A., Lehmann, U., Kelekar, A.: Autophagy delays apoptotic death in breast cancer cells following dna damage. Cell Death Differ. **14**(3), 500–510 (2006)
2. Albeck, J.G., Burke, J.M., Aldridge, B.B., Zhang, M., Lauffenburger, D.A., Sorger, P.K.: Quantitative analysis of pathways controlling extrinsic apoptosis in single cells. Mol. Cell **30**(1), 11–25 (2008)
3. Albeck, J.G., Burke, J.M., Spencer, S.L., Lauffenburger, D.A., Sorger, P.K.: Modeling a snap-action, variable-delay switch controlling extrinsic cell death. PLoS Biol. **6**(12), e299 (2008)
4. Alborzinia, H., Can, S., Holenya, P., Scholl, C., Lederer, E., Kitanovic, I., Wölfl, S.: Real-time monitoring of cisplatin-induced cell death. PloS One **6**(5), e19714 (2011)
5. Aldridge, B.B., Gaudet, S., Lauffenburger, D.A., Sorger, P.K.: Lyapunov exponents and phase diagrams reveal multi-factorial control over trail-induced apoptosis. Mol. Syst. Biol. **7**(1), 553 (2011)
6. Amaravadi, R.K., Yu, D., Lum, J.J., Bui, T., Christophorou, M.A., Evan, G.I., Thomas-Tikhonenko, A., Thompson, C.B., et al.: Autophagy inhibition enhances therapy-induced apoptosis in a myc-induced model of lymphoma. J. Clin. Invest. **117**(2), 326–336 (2007)
7. Behrends, C., Sowa, M.E., Gygi, S.P., Harper, J.W.: Network organization of the human autophagy system. Nature **466**(7302), 68–76 (2010)
8. Chen, C., Cui, J., Lu, H., Wang, R., Zhang, S., Shen, P.: Modeling of the role of a bax-activation switch in the mitochondrial apoptosis decision. Biophys. J. **92**(12), 4304–4315 (2007)
9. Crighton, D., Wilkinson, S., O'Prey, J., Syed, N., Smith, P., Harrison, P.R., Gasco, M., Garrone, O., Crook, T., Ryan, K.M.: Dram, a P53-induced modulator of autophagy, is critical for apoptosis. Cell **126**(1), 121–134 (2006)
10. Fu, Y., Glaros, T., Zhu, M., Wang, P., Wu, Z., Tyson, J.J., Li, L., Xing, J.: Network topologies and dynamics leading to endotoxin tolerance and priming in innate immune cells. PLoS Comput. Biol. **8**(5), e1002526 (2012)
11. Gaddameedhi, S., Chatterjee, S., et al.: Association between the unfolded protein response, induced by 2-deoxyglucose, and hypersensitivity to cisplatin: a mechanistic study employing molecular genomics. J. Cancer Res. Ther. **5**(9), 61 (2009)
12. Glass, L.: Combinatorial and topological methods in nonlinear chemical kinetics. J. Chem. Phys. **63**(4), 1325 (1975)
13. Gutenkunst, R.N., Waterfall, J.J., Casey, F.P., Brown, K.S., Myers, C.R., Sethna, J.P.: Universally sloppy parameter sensitivities in systems biology models. PLoS Comput. Biol. **3**(10), e189 (2007)
14. Harhaji-Trajkovic, L., Vilimanovich, U., Kravic-Stevovic, T., Bumbasirevic, V., Trajkovic, V.: Ampk-mediated autophagy inhibits apoptosis in cisplatin-treated tumour cells. J. Cell. Mol. Med. **13**(9b), 3644–3654 (2009)
15. Heppner, G.H.: Tumor heterogeneity. Cancer Res. **44**(6), 2259–2265 (1984)
16. Hong, T., Xing, J., Li, L., Tyson, J.J.: A mathematical model for the reciprocal differentiation of t helper 17 cells and induced regulatory t cells. PLoS Comput. Biol. **7**(7), e1002122 (2011)
17. Huett, A., Goel, G., Xavier, R.J.: A systems biology viewpoint on autophagy in health and disease. Curr. Opin. Gastroenterol. **26**(4), 302–309 (2010)
18. Jegga, A.G., Schneider, L., Ouyang, X., Zhang, J.: Systems biology of the autophagy-lysosomal pathway. Autophagy **7**(5), 477–489 (2011)
19. Kapuy, O., Vinod, P.K., Mandl, J., Bánhegyi, G.: A cellular stress-directed bistable switch controls the crosstalk between autophagy and apoptosis. Mol. Biosyst. **9**(2), 296–306 (2013)
20. Kimmelman, A.C.: The dynamic nature of autophagy in cancer. Genes Dev. **25**(19), 1999–2010 (2011)
21. Komatsu, M., Ueno, T., Waguri, S., Uchiyama, Y., Kominami, E., Tanaka, K.: Constitutive autophagy: vital role in clearance of unfavorable proteins in neurons. Cell Death Differ. **14**(5), 887–894 (2007)

22. Loewer, A., Lahav, G.: We are all individuals: causes and consequences of non-genetic heterogeneity in mammalian cells. Curr. Opin. Genet. Dev. **21**(6), 753–758 (2011)
23. Maiuri, M.C., Zalckvar, E., Kimchi, A., Kroemer, G.: Self-eating and self-killing: crosstalk between autophagy and apoptosis. Nat. Rev. Mol. Cell Biol. **8**(9), 741–752 (2007)
24. Martin, K.R., Barua, D., Kauffman, A.L., Westrate, L.M., Posner, R.G., Hlavacek, W.S., MacKeigan, J.P.: Computational model for autophagic vesicle dynamics in single cells. Autophagy **9**(1), 53–52 (2013)
25. Marusyk, A., Polyak, K.: Tumor heterogeneity: causes and consequences. Biochim. Biophys. Acta (BBA)-Rev. Cancer. **1805**(1), 105–117 (2010)
26. Michie, A.M., McCaig, A.M., Nakagawa, R., Vukovic, M.: Death-associated protein kinase (DAPK) and signal transduction: regulation in cancer. FEBS J. **277**(1), 74–80 (2010)
27. Mjolsness, E., Sharp, D.H., Reinitz, J.: A connectionist model of development. J. Theor. Biol. **152**(4), 429–453 (1991)
28. Musiwaro, P., Smith, M., Manifava, M., Walker, S.A., Ktistakis, N.T.: Characteristics and requirements of basal autophagy in HEK 293 cells. Autophagy **9**(9), 1407–1417 (2013)
29. Nakai, A., Yamaguchi, O., Takeda, T., Higuchi, Y., Hikoso, S., Taniike, M., Omiya, S., Mizote, I., Matsumura, Y., Asahi, M., et al.: The role of autophagy in cardiomyocytes in the basal state and in response to hemodynamic stress. Nat. Med. **13**(5), 619–624 (2007)
30. Ogata, M., Hino, S.-I., Saito, A., Morikawa, K., Kondo, S., Kanemoto, S., Murakami, T., Taniguchi, M., Tanii, I., Yoshinaga, K., et al.: Autophagy is activated for cell survival after endoplasmic reticulum stress. Mol. Cell. Biol. **26**(24), 9220–9231 (2006)
31. Perez, R.P.: Cellular and molecular determinants of cisplatin resistance. Eur. J. Cancer **34**(10), 1535–1542 (1998)
32. Periyasamy-Thandavan, S., Jiang, M., Wei, Q., Smith, R., Yin, X.-M., Dong, Z.: Autophagy is cytoprotective during cisplatin injury of renal proximal tubular cells. Kidney Int. **74**(5), 631–640 (2008)
33. Pinto, D.J., Brumberg, J.C., Simons, D.J., Ermentrout, G.B., Traub, R.: A quantitative population model of whisker barrels: re-examining the Wilson-Cowan equations. J. Comput. Neurosci. **3**(3), 247–264 (1996)
34. Rehm, M., Huber, H.J., Dussmann, H., Prehn, J.H.M.: Systems analysis of effector caspase activation and its control by x-linked inhibitor of apoptosis protein. EMBO J. **25**(18), 4338–4349 (2006)
35. Rehm, M., Huber, H.J., Hellwig, C.T., Anguissola, S., Dussmann, H., Prehn, J.H.M.: Dynamics of outer mitochondrial membrane permeabilization during apoptosis. Cell Death Differ. **16**(4), 613–623 (2009)
36. Rodriguez-Rocha, H., Garcia-Garcia, A., Panayiotidis, M.I., Franco, R.: DNA damage and autophagy. Mutat. Res. Fundam. Mol. Mech. Mutagen. **711**(1), 158–166 (2011)
37. Rosenberg, B., Vancamp, L.: Platinum compounds: a new class of potent antitumour agents. Nature **222**, 385–386 (1969)
38. Sanchez-Perez, I., Murguia, J.R., Perona, R.: Cisplatin induces a persistent activation of jnk that is related to cell death. Oncogene **16**(4), 533 (1998)
39. Scarlatti, F., Granata, R., Meijer, A.J., Codogno, P.: Does autophagy have a license to kill mammalian cells? Cell Death Differ. **16**(1), 12–20 (2008)
40. Singhania, R.: Modeling protein regulatory networks that control mammalian cell cycle progression and that exhibit near-perfect adaptive responses. Ph.D. thesis (2011)
41. Spencer, S.L., Sorger, P.K.: Measuring and modeling apoptosis in single cells. Cell **144**(6), 926–939 (2011)
42. Spencer, S.L., Gaudet, S., Albeck, J.G., Burke, J.M., Sorger, P.K.: Non-genetic origins of cell-to-cell variability in trail-induced apoptosis. Nature **459**(7245), 428–432 (2009)
43. Tyson, J.J., Novák, B.: Functional motifs in biochemical reaction networks. Annu. Rev. Phys. Chem. **61**, 219–240 (2010)
44. Tyson, J.J., Baumann, W.T., Chen, C., Verdugo, A., Tavassoly, I., Wang, Y., Weiner, L.M., Clarke, R.: Dynamic modelling of oestrogen signalling and cell fate in breast cancer cells. Nat. Rev. Cancer **11**(7), 523–532 (2011)

45. Wei, Y., Pattingre, S., Sinha, S., Bassik, M., Levine, B.: JNK1-mediated phosphorylation of BCL-2 regulates starvation-induced autophagy. Mol. Cell **30**(6), 678–688 (2008)
46. Wei, Y., Sinha, S.C., Levine, B.: Dual role of JNK1-mediated phosphorylation of BCL-2 in autophagy and apoptosis regulation. Autophagy **4**(7), 949–951 (2008)
47. White, E.: Autophagic cell death unraveled: pharmacological inhibition of apoptosis and autophagy enables necrosis. Autophagy **4**(4), 399–401 (2008)
48. Wilson, H.R., Cowan, J.D.: Excitatory and inhibitory interactions in localized populations of model neurons. Biophys. J. **12**(1), 1–24 (1972)
49. Wu, J., Hu, C.-P., Gu, Q.-H., Li, Y.-P., Song, M.: Trichostatin a sensitizes cisplatin-resistant A549 cells to apoptosis by up-regulating death-associated protein kinase. Acta Pharmacol. Sin. **31**(1), 93–101 (2010)
50. Xu, Y., Yuan, J., Lipinski, M.M.: Live imaging and single-cell analysis reveal differential dynamics of autophagy and apoptosis. Autophagy **9**(9), 1418–1430 (2013)
51. Yang, Z., Klionsky, D.J.: Eaten alive: a history of macroautophagy. Nat. Cell Biol. **12**(9), 814–822 (2010)
52. Yuan, T.L., Wulf, G., Burga, L., Cantley, L.C.: Cell-to-cell variability in PI3K protein level regulates PI3L-AKT pathway activity in cell populations. Curr. Biol. **21**(3), 173–183 (2011)
53. Zalckvar, E., Berissi, H., Eisenstein, M., Kimchi, A.: Phosphorylation of beclin 1 by dap-kinase promotes autophagy by weakening its interactions with BCL-2 and BCL-XL. Autophagy **5**(5), 720–722 (2009)
54. Zalckvar, E., Berissi, H., Mizrachy, L., Idelchuk, Y., Koren, I., Eisenstein, M., Sabanay, H., Pinkas-Kramarski, R., Kimchi, A.: Dap-kinase-mediated phosphorylation on the BH3 domain of beclin 1 promotes dissociation of beclin 1 from BCL-XL and induction of autophagy. EMBO Rep. **10**(3), 285–292 (2009)
55. Zalckvar, E., Yosef, N., Reef, S., Ber, Y., Rubinstein, A.D., Mor, I., Sharan, R., Ruppin, E., Kimchi, A.: A systems level strategy for analyzing the cell death network: implication in exploring the apoptosis/autophagy connection. Cell Death Differ. **17**(8), 1244–1253 (2010)
56. Zhang, T., Brazhnik, P., Tyson, J.J.: Computational analysis of dynamical responses to the intrinsic pathway of programmed cell death. Biophys. J. **97**(2), 415–434 (2009)

# Chapter 3
# An Experimental Framework to Study the Dynamics of Autophagic Response

In this chapter we present an experimental quantitative framework for measuring kinetic parameters such as autophagy flux, time course of autophagic response, and stress/response dynamics in single cancer cells including endocrine-resistant breast cancer cells. Some primary data are presented in this chapter. All the microscopic images were collected using a Nikon Eclipse TE-300 Spinning Disk Time-lapse Microscope System. Data collection was done using Velocity 3D Image Analysis Software. Image analysis was done by ImageJ software [1].

Xu et al. [55] have developed a methodology using automated live microscopy to study the time course of autophagy and apoptosis. They have used quantitative single-cell imaging in H4 glioma cells, L929 mouse connective tissue cells, and HeLa human adenocarcinoma cells to study interplay of autophagy and apoptosis. In their work, they used starvation, rapamycin (an inhibitor of mTOR), and staurosporine (STS) as the stressors to induce autophagy and apoptosis in these cells [55]. The methodology they presented can be used to validate the mathematical model presented in Chap. 2 and is very useful to capture the sress/reponse plots and time course of autophagy induction and progression.

Although the framework presented here is for detecting dynamics of interplay of autophagy and apoptosis in single cancer cells as proposed by Xu et al. [55], our goal is to address the roadmap necessary to study autophagy dynamics and cell fate in endocrine-resistant breast cancer cells and to compare the dynamics of autophagic response in sensitive and resistant breast cancer cells to therapeutic regimens including endocrine therapy.

This methodology and also the methodology proposed and validated by Xu et al. [55] are the first steps to study the dynamics of autophagy and its crosstalk with other cellular processes such as apoptosis, necrosis, cell cycle, and cell growth. These methodologies will give us quantitative information about the inputs (stress type, stress strength, stress time interval) and the outputs (autophagy, cell death, cell survival, quiescence) in cells controlled by signaling network of interaction of autophagy and apoptosis. This information will be crucial to validating dynamic



I. Tavassoly, *Dynamics of Cell Fate Decision Mediated by the Interplay of Autophagy and Apoptosis in Cancer Cells*, Springer Theses, DOI 10.1007/978-3-319-14962-2_3

models of the signaling networks. These methods will have to be supplemented by other methods to track dynamics of other components of the signaling networks of interplay of autophagy and apoptosis such as JNK expression, DAPK expression, cytoplasmic calcium, IP3R activity, and BCL2 proteins expression in ER and mitochondrion and mTOR activation and inhibition.

With the data provided by methodology developed by Xu et al. [55] and the framework presented here, it is not possible to estimate directly the parameters presented in Table 2.4, but it is possible to propose parameter sets that model reproduces the observed dynamics for stress/response and time course of autophagosome formation.

## 3.1 Methods to Detect and Measure Autophagy

Apoptosis has been studied extensively and there are several efficient methods for detecting it, including annexin V/propidium iodide, TUNEL assay, and cell morphology [24, 27, 44, 45, 51]. Dynamics of apoptosis also have been studied experimentally, and time-lapse imaging of cytochrome c release has shown the kinetic of apoptosis initiation [19–21]. The on/off switch of apoptosis has been studied experimentally as well [47].

The first method used to detect autophagy was Transmission Electron Microscopy (TEM) [15, 30]. This method can reveal the structure of autophagosomes structure and their contents but the data provided by TEM are static and from dead cells, and it seems very difficult to capture time course data by this method. Autophagic vacuoles were first recognized in the kidney cells of newborn mice in 1957 by electron microscopy [10, 53]. Ashford and Porter [3] described the morphology of autophagosome using electron microscopy in rat liver cells. In fact the term "AUTOPHAGY" seems to be used first in 1967 by Deter et al. [13] in an electron microscopic study of rat liver cells [53]. Interestingly they measured size distribution of autophagic vacuoles quantitatively [13].

Live-cell imaging using fluorescence microscopy is a powerful method for tracking autophagy. Microtubule-associated protein light chain 3 (LC3), a mammalian homologue of Saccharomyces cerevisiae Atg8, is one of the proteins participating in formation of autophagosomes. LC3-II, the membrane-bound form of LC3, is formed during autophagosome formation and has been proposed and widely used as a marker for autophagy. Tracking autophagy markers (such as GFP-LC3) can be a good tool to measure dynamic aspects of this cellular response [29, 37].

A method for detecting autophagy has been proposed by Katayama et al. [28] which uses Kaima as a autophagy marker. Kaima is a acid-stable fluorescent probe which is able to show the formation of autolysosomes and can be used to track autophagy in live cells. GFP-LC3 cannot detect autolysosomes because, after docking of autophagosomes and lysosomes, LC3-II on outer membrane of autophagosomes turns into LC3-I via delipidation, and this LC3-I will go back to the cytoplasmic pool of LC3-I to be reused. LC3-II on the inner membrane of autophagosomes is degraded in autolysosomes. Kaima-based probes are stable in acidic pH and can remain stable in autolysosomes, therefore they can be used to detect the process of autolysosome formation and degradation of autophagosomes contents [28, 41].

Although LC3-II shows the formation and accumulation of autophagosomes, it has been shown that a fraction of LC3-II proteins are degraded by lysosomal enzymes [48].

SQSTM1/P62 is another marker for autophagic activity in cells. SQSTM1 takes part in formation of autophagosomes, and it is degraded in autolysosomes; in fact, it is a marker for autophagosome degradation. Inhibition of the docking of autophagosomes with lysosomes leads to increased concentration of SQSTM1 in the cytoplasm, and SQSTM1 level decreases dramatically during degradation of autophagosomes [29, 40]. To study the autophagy flux in cells, it is necessary to use both LC3 and SQSTM1 as autophagy markers, SQSTM1 level is always diminished when LC3-II level is increased [29].

Bampton et al. [4] have compared different markers for monitoring autophagy in live cells. They found that EGFP-LC3 can be used to detect autophagosomes in live cells in different stages of autophagy prior to formation of autolysosomes. To have an accurate monitoring system for tracking autophagosome formation and degradation they have indicated the importance of using other markers for autolysosomes and autophagosomes degradation [4].

Quantification of GFP-LC3 puncta is a method to measure the level and extent of autophagic activity, especially formation of autophagosomes and their accumulation [29, 30]. Although there are many concerns about accuracy and efficiency of quantification of GFP-LC3 puncta, this marker is still a useful probe for monitoring autophagic response, especially in live cells. Different methods can be applied for quantification of GFP-LC3 puncta. The very first method is simply to count the number of cells with positive GFP-LC3 puncta. This method is useful when quantitative aspects of autophagy is being studied in a population of cells. Another way to quantify GFP-LC3 is to count the number of GFP-LC3 puncta per cell [30]. It seems that this method is not very applicable for tracking autophagy in single live cells because usually the extent (area) of GFP-LC3 puncta rather than number of GFP-LC3 is increased in a single cancer cell in stress condition such as starvation or drug treatment (Figs. 3.21 and 3.23). Another alternative is to measure the area of GFP-LC3 puncta per cell, although there is no clear evidence how to connect the number or area of GFP-LC3 puncta to quantitative characteristics of autophagosomes. This method has been used for quantification of autophagic process [39], and we have used it for quantification of autophagic response in human H4 cells and breast cancer cells. We have observed that both GFP-LC3 puncta area and number increase when cancer cells are under stress.

Flow cytometry and Fluorescence Activated Cell Sorter (FACS) are other tools for monitoring quantitative aspects of autophagy in cells using GFP-LC3 [46].

Cell cycle phase is a determinant factor affecting the dynamics of autophagic response [18, 43]. A new strategy for tracking autophagy in different phases of the cell cycle has been developed by Kaminskyy et al. [27]. In this method, which is an automated flow cytometry-based assay, membrane unbound LC3 proteins (LC3-I) is extracted from cells. Then the remaining LC3-II proteins are stained as markers of autophagosomes. Staining with propidium iodide is used for tracking of autophagosome accumulation in different phases of the cell cycle [27]. A method for capturing the dynamics of protein expression during the cell cycle using fixed cells has been introduced which seems to be a good method to study the dynamics of autophagy in cancer cells using proper markers such as LC3 and SQSTM1 [26].

Although dynamic properties of apoptosis have been studied, there are limited quantitative data on autophagy time courses and cell fate events during autophagic response especially in breast cancer cells. Because of heterogeneity of tumors and cell lines, the first step to study temporal functions of cellular stress responses is to use single cells as model systems [34].

## 3.2 Quantitative Parameters of Autophagic Response

As the mathematical model presented in Chap. 2 has implied, there are a lot of parameters related to autophagy and apoptosis dynamics which determine the cell fate after different stresses. To validate the model predictions and find more accurate parameters for the model, a lot of parameters need to be measured quantitative. The model presented in Chap. 2 can be parameterized for different types of cell lines including different cancer cell lines. As measuring all the parameters in the model is very difficult, the first step can be measuring the stress/response, time course of autophagy, time-points when apoptosis occurs and percentage of apoptosis in cell populations.

Measuring the time course of protein expressions including expression of JNK and DAPK is one of the basic steps to assess autophagic response in time.

Loos and Engelbrecht [35] have suggested measuring a series of kinetic parameters related to cell death modes to evaluate the dynamics of their interplay. Parameters related to autophagic response include: basal autophagy, maximum autophagic capacity, time delay between induction of stress and initiation of autophagy, time delay to reach maximum autophagy level, and kinetic parameters related to ATP synthesis and hydrolysis [35].

All of these parameters help to understand autophagy as a dynamical response at a phenomenological level. For modeling purposes we can add the parameters shown in Table 2.4. As mentioned before designing experimental settings to be able to give us the opportunity to directly measure the parameters used in Wilson–Cowan equations of protein interactions has not been studied yet but phenomenological quantitative data can help us to estimate them optimally.

Kinetic parameters related to apoptotic cell death include time delay between induction of stress and initiation of apoptosis at different levels (cytochrome c release, initiation of MOMP, caspase-3 cleavage), time delay to express different stages of apoptosis morphology such as chromatin condensation (pyknosis) and p53 localization [35].

In the roadmap designed by Loos and Engelbrecht [35] other parameters related to necrosis kinetics, intracellular metabolism, extracellular metabolism, and stress parameters have been mentioned. For necrosis some of these parameters can be summarized as time needed to induce membrane leakage, time delay to display lysosomal leakage, and time course of loss of membrane integrity [35]. There are some cell specific parameters such as mitochondrial load, BCL2 protein distribution in different compartments, cell state (proliferating cell, senescent cell, quiescent cell), and subcellular organelle properties [35]. Stress parameters can be stress type, stress duration, stress strength, and mechanism of stress induction and action [35].

## 3.3 Experimental Observations in Human H4 Neuroglioma Cells

H4 cells stably expressing GFP-LC3 and H2B-RFP (LC3-H2B Cells) were obtained from Xu et al. [55]. Here Histone H2B-RFP was used to detect the initiation of apoptosis, because apoptosis involves posttranslational histone modifications including phosphorylation of histone H2B [9]. H2B-RFP has been used as a marker for tracking nucleus and cell identification in live cell imaging [55]. Condensation and cleavage of DNA are among the main characteristics of apoptotic cell death, and labeling DNA indirectly by histone-fluorescent protein fusion can be a method to study the time course of apoptosis in live cells [55].

Figure 3.1 shows basal autophagy in two H4 neuroglioma cells expressing GFP-LC3 and H2B-RFP (left panel shows H2B-RFP, right panel shows GFP-LC3 and the middle panel is the merged image). Movies 1.1, 1.2, 1.3 are real-time microscopy of these cells (Movie 1.1 shows GFP-LC3, Movie 1.2 shows H2B-RFP, and Movie 1.3 is merged format of Movie 1.1 and 1.2).

Constitutive autophagy is a basic feature observed in cancer cells expressing GFP-LC3, when there is no stress. The mechanism behind the heterogeneity in basal autophagy among cancer cells is an important question to be answered (Fig. 3.2).

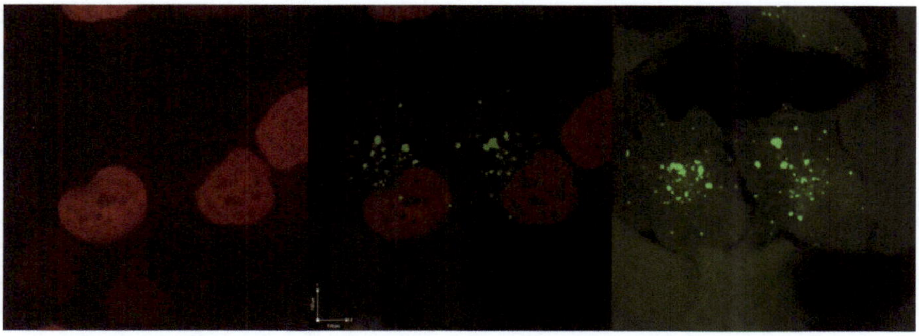

**Fig. 3.1** Human H4 neuroglioma cells stably expressing GFP-LC3 and H2B-RFP

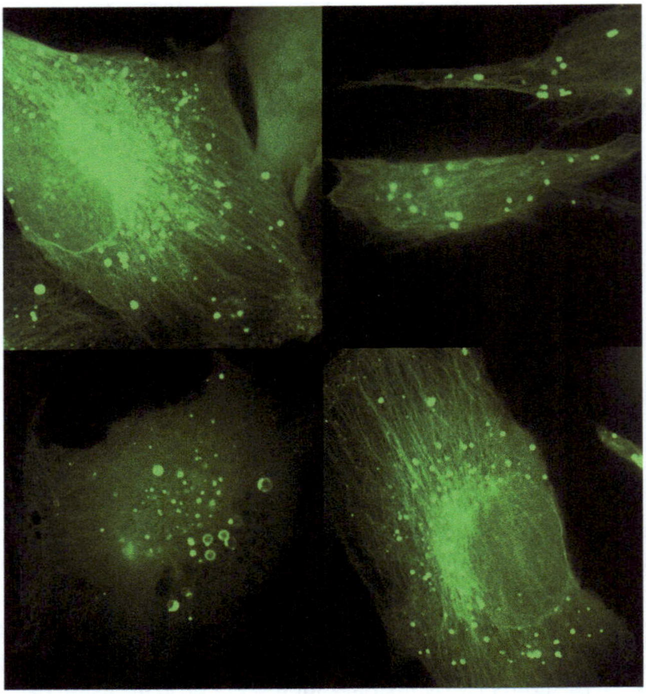

**Fig. 3.2** Heterogeneity in basal autophagy represented by GFP-LC3 puncta in different H4 neuroglioma cells

One very basic answer can be the general heterogeneity of proteins expression level in cancer cells. Movies 2, 3, and 4 are real-time microscopy of basal autophagy marked by GFP-LC3 puncta in single H4 neuroglioma cells. In these real-time microscopies, movements of GFP-LC3 puncta are observed.

Figure 3.3 shows the different distribution of GFP-LC3 puncta in different H4 cells. The other interesting feature of basal autophagy in cancer cells is different patterns of distribution of autophagosome expression in cytoplasm, marked by GFP-LC3. Which parameters determine the accumulation of autophagosomes in different parts of cytoplasm is another unanswered question. For approaching this question, we need to use fluorescent markers for different autophagy targets and track them during the autophagy process. Considering selective autophagy, autophagosomes may accumulate in different locations of cytoplasm to degrade specific targets (mitophagy, ER-phagy, glycophagy, ribophagy, etc.).

Movie 5 and Fig. 3.4 show the time-lapse microscopy of basal autophagy in a single H4 neuroglioma cell for 1,800 min. This cell keeps its basal level of autophagy during this time, although some fluctuations are observed due to changes in cell environment such as decreased nutrients or oxygen. In movie 5, images have been taken every 15 min.

To capture the dynamics of autophagy and apoptosis in H4 neuroglioma cells, we need to find the time course of increased autophagy level (GFP-LC3 puncta) with

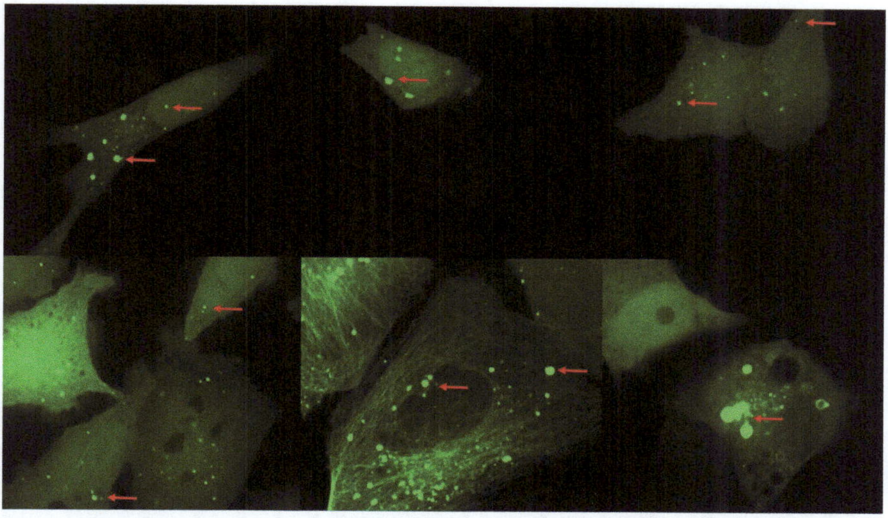

**Fig. 3.3** Heterogeneity in GFP-LC3 puncta distribution pattern in H4 neuroglioma cells (basal autophagy)

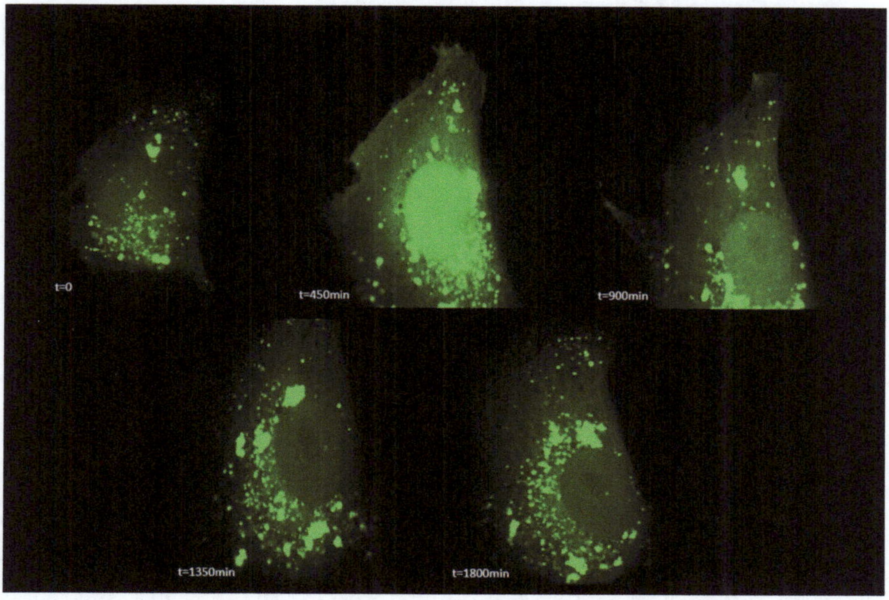


**Fig. 3.4** Basal autophagy in a H4 cell

different stresses (different stress types, strengths, and durations). We have shown that cisplatin treatment can be a good stressor for this purpose. Figures 3.5 and 3.6 are time-lapse microscopy of two H4 neuroglioma cells (one expressing GFP-LC3 and one expressing both GFP-LC3 and H2B-RFP) treated with 1 mM of cisplatin.

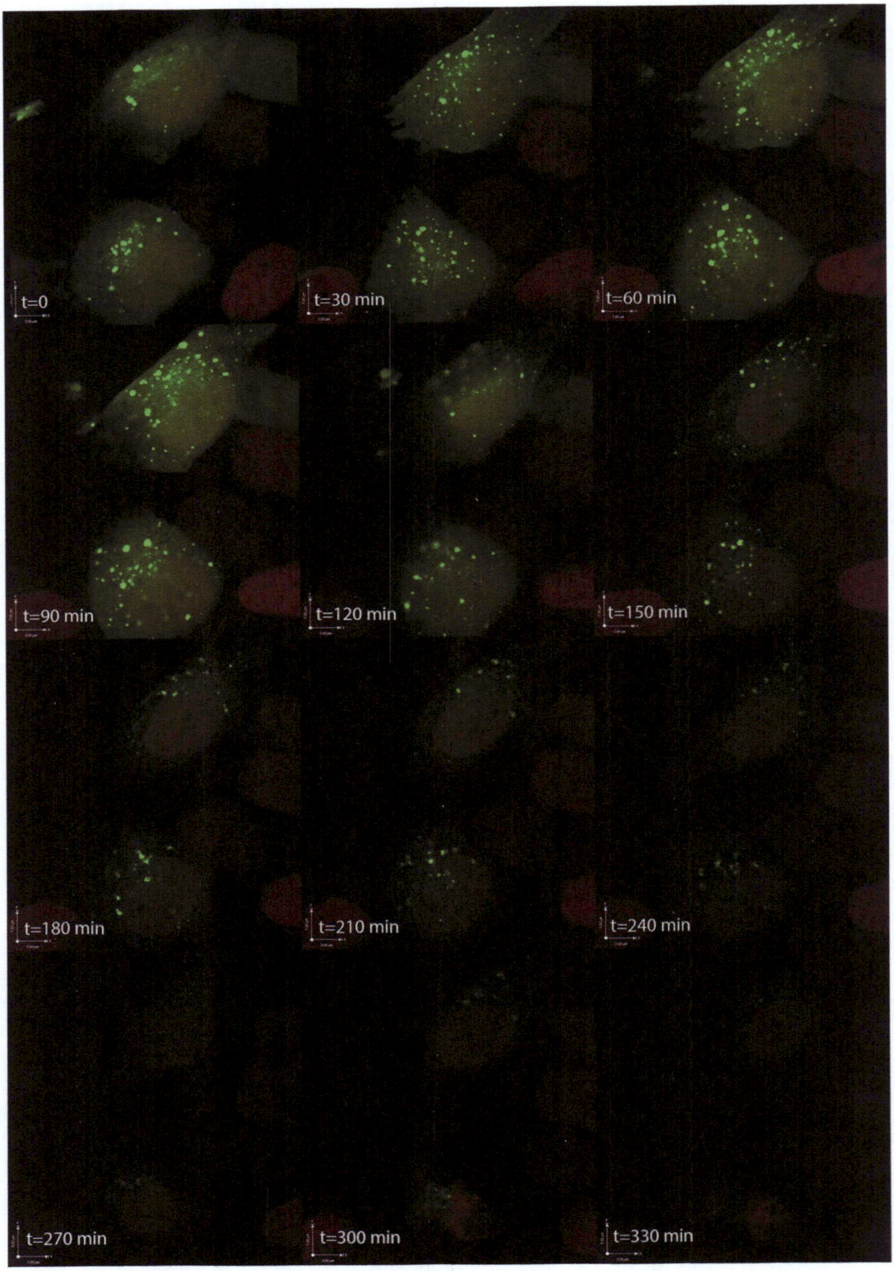


**Fig. 3.5** Time-lapse microscopy of a H4 cell expressing GFP-LC3/H2B-RFP treated with 1 mM cisplatin

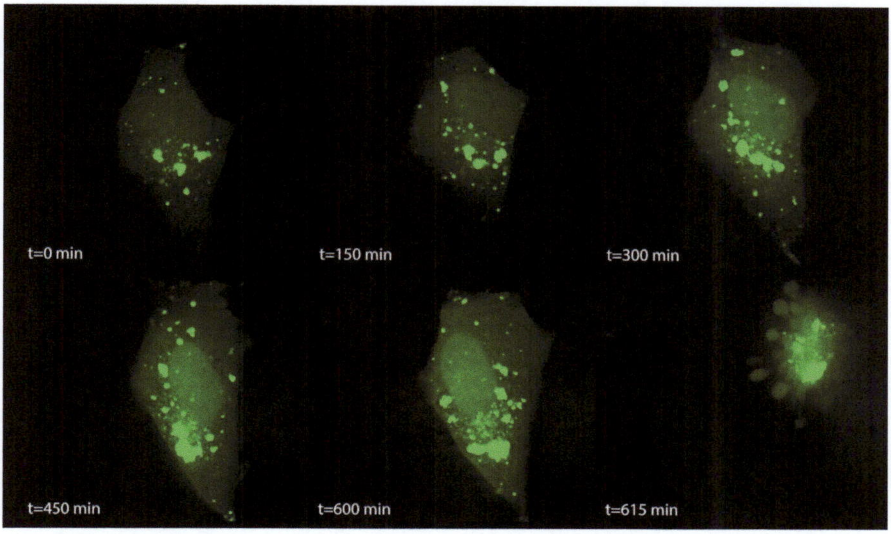


**Fig. 3.6** Time-lapse microscopy of autophagy in a H4 cell expressing GFP-LC3 treated with 1 mM cisplatin

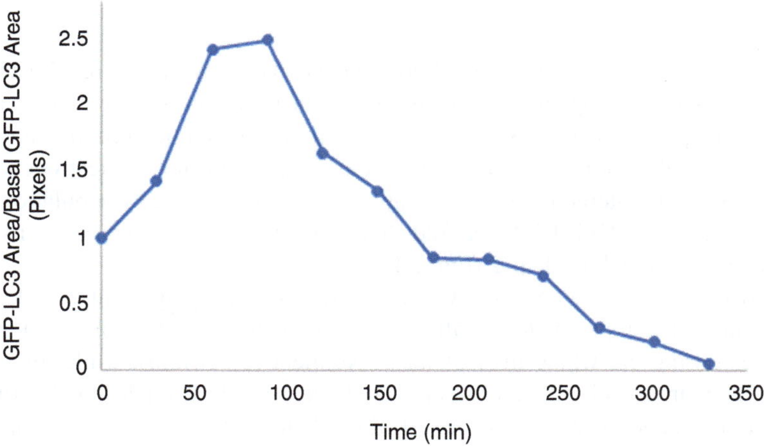


**Fig. 3.7** Time course of GFP-LC3 puncta area fold change in a single H4 cell treated with 1 mM cisplatin

Previous study has shown that autophagy is induced in a time- and dose-dependent pattern in glioma cells [23].

Movies 6.1, 6.2, and 6.3 are time-lapse microscopy of the H4 cell shown in Fig. 3.5 (GFP-LC3 channel, H2B-RFP channel, and merged channel). Movie 7 is time-lapse microscopy of the H4 cell shown in Fig. 3.6. For the first case cell images were taken every 30 min and for the second one cell images were taken every 15 min.

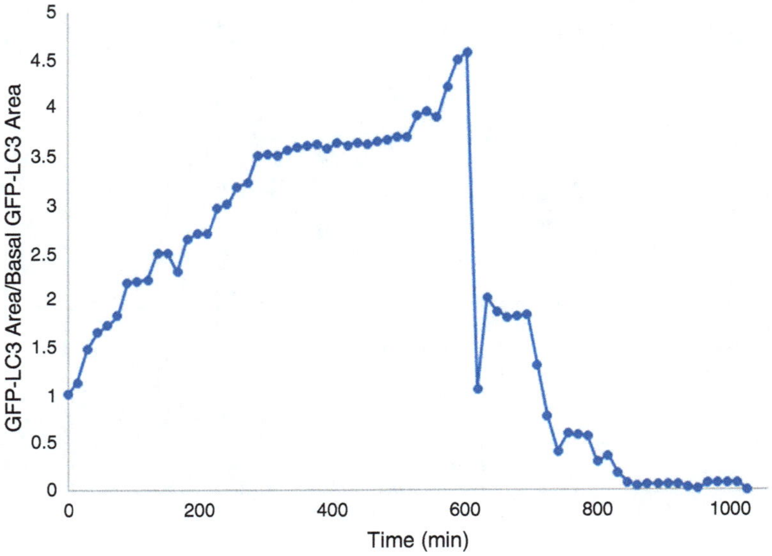


**Fig. 3.8** Time course of GFP-LC3 puncta area fold change in a single H4 cell expressing GFP-LC3 treated with 1 mM cisplatin

Quantification of GFP-LC3 area fold change in these two cells (Figs. 3.5 and 3.6) shows increased autophagy and then cell death by apoptosis after 300 min and 615 min, respectively (Figs. 3.7 and 3.8). The difference in time to commit apoptosis after the same treatment in these two cells is an indicator of heterogeneity in dynamics of the interaction of autophagy and apoptosis at a cell population level. But the patterns of GFP-LC3 area fold change are consistent with predictions of the mathematical model in Chap. 2 (Fig. 2.4).

Movies 8.1, 8.2, and 8.3 show a single H4 cell escaping from cell death by increasing its autophagy level, after treatment with 50 μM cisplatin. This cell survives during the whole time of the experiment (1,320 min), while three other cells which are seen in the microscopic field commit cell death during this time. In this movie, images were taken every 30 min and due to the cell movement out of the microscopic focus field, quantification of GFP-LC3 was not possible.

## 3.4 Experimental Observations in Breast Cancer Cells

A series of cell lines have been derived from MCF7 cells to study estrogen-independence and antiestrogen resistance in breast cancer. MCF7 cells are estrogen-dependent and sensitive to tamoxifen and fulvestrant. Tamoxifen and Fulvestrant (Faslodex, ICI 182,780) are estrogen receptor antagonists which are used in treatment of estrogen receptor-positive breast tumors [11, 50]. LCC1 cells are a variant

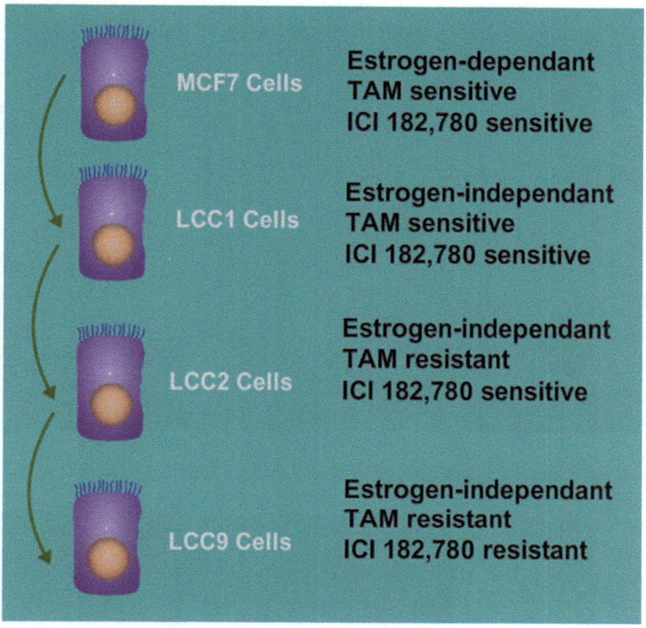


**Fig. 3.9** Evolution and characteristics of endocrine-resistant breast cancer cell lines

of MCF7 cells which are estrogen-independent but they are sensitive to tamoxifen and fulvestrant. LCC2 cells are estrogen-independent, resistant to tamoxifen and sensitive to fulvestrant. LCC9 cells are estrogen-independent and resistant to both tamoxifen and fulvestrant [6, 7, 11, 12, 50]. Figure 3.9 has summarized the evolution of these breast cancer cell lines and their characteristics (this figure was made by online protein lounge tools).

For tracking autophagy in these breast cancer cells, MCF7, LCC1, and LCC9 were transiently transfected with GFP-LC3 as a marker for autophagy (Figs. 3.10, 3.11, and 3.12).

### *3.4.1 Basal Autophagy in Breast Cancer Cells*

MCF7, LCC1, and LCC9 cells keep a constitutive level of autophagy. There is no report in literature to compare basal autophagy in sensitive and resistant breast cancer cells to endocrine therapy. The main difficulties in comparing basal autophagy in these cell lines are the wide-ranging heterogeneity in cell size and the variable basal level of autophagy even within each of these cell lines (Fig. 3.13).

We randomly selected 100 cells from MCF7 and 100 cells from LCC9 cells, while they were expressing GFP-LC3, and quantified the GFP-LC3 puncta area in each group. We calculated the average level of basal GFP-LC3 puncta area in these

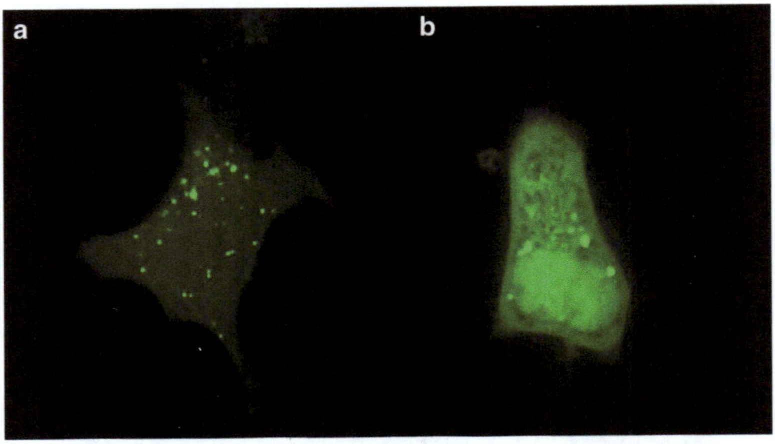


**Fig. 3.10** Basal autophagy in an MCF7 cell (**a**) and an LCC9 cell (**b**) expressing GFP-LC3

**Fig. 3.11** Basal autophagy in LCC1 cells characterized by GFP-LC3 puncta

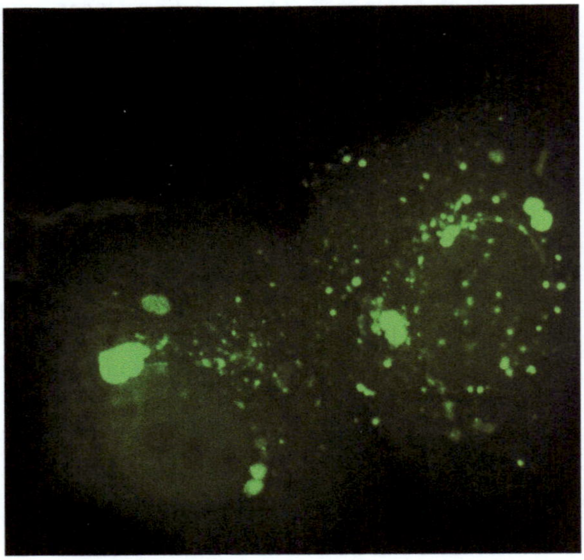

cell lines. Although this method is not very accurate, but at least it can help to estimate basal autophagy at level of cell population and compare these two cell lines. All images were taken by 100× magnification and are shown in Figs. 3.14 and 3.15. Our basic calculation shows that in LCC9 cells average GFP-LC3 puncta area is 1,079.33 (pixel) and in MCF7 cells it is 607.85 (pixel). Even qualitatively looking at a population of LCC9 cells shows a higher level of autophagy compared to MCF7 cells. More experimental validation is needed to see why and how resistant cancer cells keep their basal autophagy higher compared to sensitive cells.

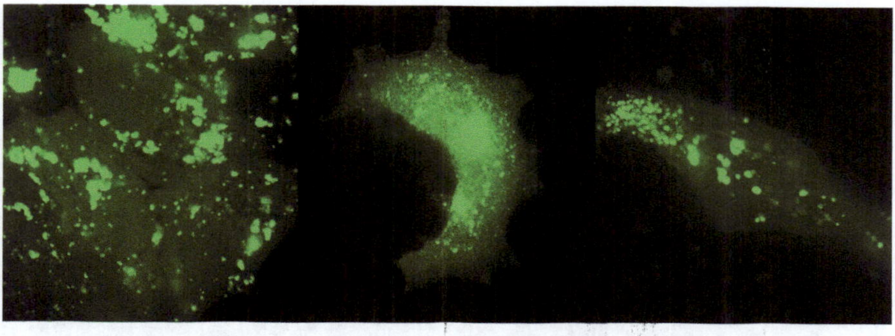

**Fig. 3.12** Basal autophagy in LCC9 cells characterized by GFP-LC3 puncta

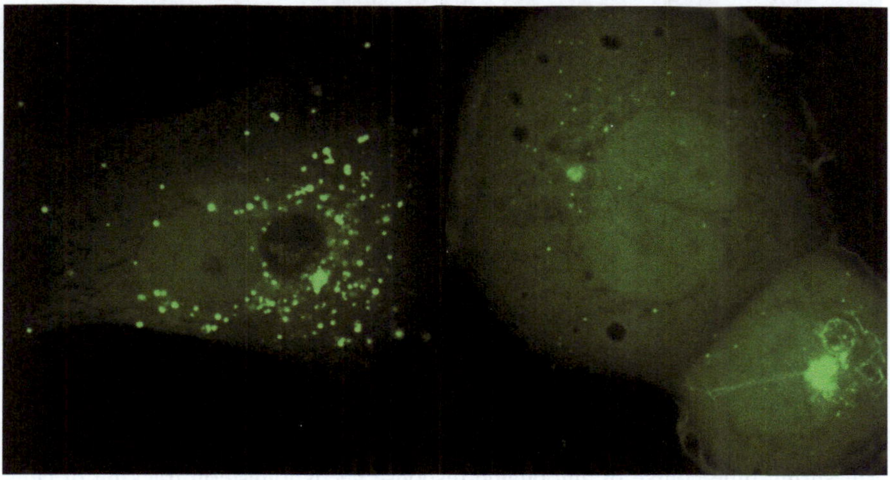

**Fig. 3.13** Basal autophagy in MCF7 cells characterized by GFP-LC3 puncta

Movies 9.1 and 9.2 are real-time movies of basal autophagy and GFP-LC3 puncta movements in MCF7 cells. Movie 10 shows real-time microscopy of basal autophagy in LCC1 cells expressing GFP-LC3 and movies 11.1, 11.2, and 11.3 are from LCC9 cells expressing GFP-LC3 as a marker of their basal autophagy.

Figure 3.16 is time-lapse microscopy of basal autophagy in a single LCC9 cell expressing GFP-LC3. This cell keeps its basal autophagy and sometimes increases it, maybe because of stresses in the environment. Movie 12 shows this cell and its basal autophagy during 600 min (images were taken every 15 min).

Movie 13 is time-lapse microscopy of another single LCC9 for 675 min (Images were taken every 15 min). Movie 14 shows tracking of basal autophagy in a single LCC1 cell expressing GFP-LC3 for 47 h (images were taken every 20 min). The time-lapse microscopy for this cell started upon plating and the movie shows how this cell regulates its basal autophagy when it arrives in a new environment after plating.

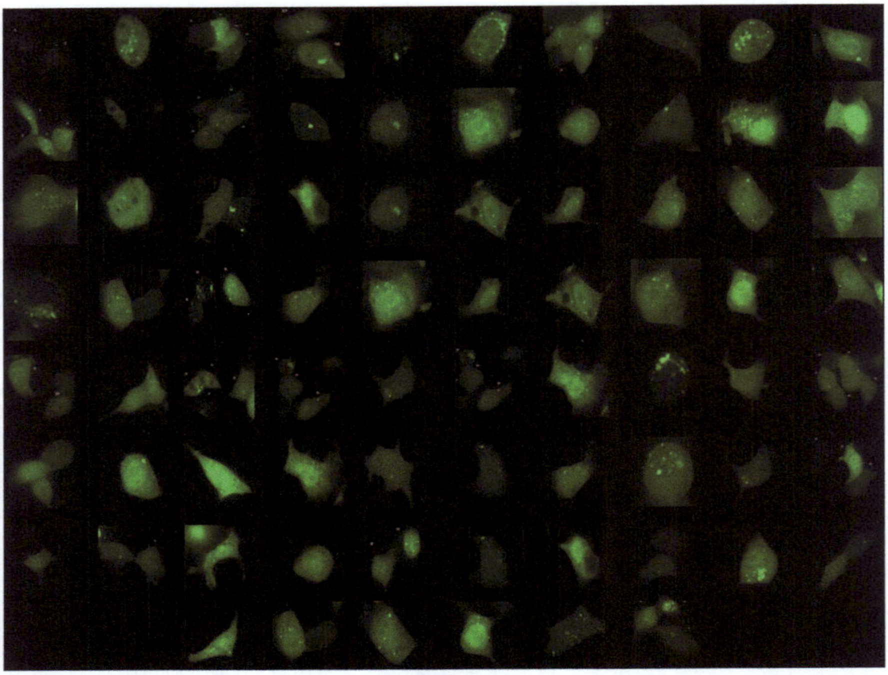

**Fig. 3.14** Basal autophagy marked by GFP-LC3 in 100 MCF7 cells

### *3.4.2 Autophagy, Cell Growth, and Cell Cycle*

Measuring time course of autophagic response needs long-term tracking of cells. During this time, cells, especially when using heterogeneous and unsynchronized population of cancer cell lines, will go through different phases of cell cycle including mitosis and cell division. It is evident that autophagy, cell cycle, and cell growth interact with each other and affect each other's dynamics [8, 17, 22, 32, 38, 43, 49, 52, 56].

Although autophagy is involved in cell growth control, exact mechanisms of the interaction of autophagy with cell growth are not understood. The role of autophagy in controlling cell growth may be defined through its interaction with PI3K/TOR signaling [38, 52].

The accurate measurement of autophagic flux in single cells needs considering the volume and size of each cell, and long-term tracking of autophagic response in single cells needs considering cell division and cell cycle. The reason is the significant changes in autophagosome formation during different phases of cell cycle, including mitosis. Investigations have found that autophagy is shut down during mitosis. It seems that cells need to protect the spindle apparatus and chromosomes from being degraded by the autophagy machinery [14]. But in some other studies persistent autophagy has been observed during mitosis. The possible

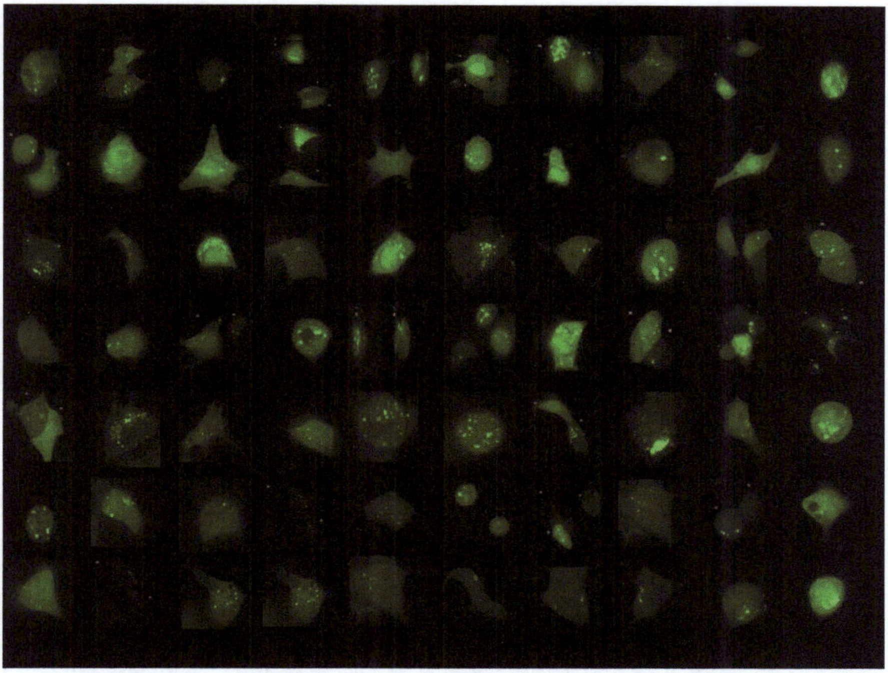

**Fig. 3.15** Basal autophagy marked by GFP-LC3 in 100 LCC9 cells

**Fig. 3.16** Time-lapse microscopy of basal autophagy in an LCC9 cell expressing GFP-LC3

hypothesis for the latter scenario is need of mitotic cells to get rid of their damaged mitochondria, because mitosis is a high energy demanding process which can cause damaged and dysfunctional mitochondria [33].

Furuya et al. [18] have shown that VPS34 (one of the main proteins taking part in Beclin-1 complex formation) is phosphorylated on Thr159 by Cdk1. This phosphorylation inhibits VPS34 interaction with Beclin-1, which means knockdown of autophagosome formation during mitosis. Cdk5/p52 can phosphorylate Vpd34 with the same mechanism [18, 54].

Kaminskyy et al. [27] have measured autophagic flux in different phases of cell cycle in cancer cells using a flow cytometry-based approach. They have measured the basal level of autophagy in G1, S, and G2/M phases of the cell cycle. Their results have shown that cells in G0/G1 phase have the highest level of basal autophagy [27].

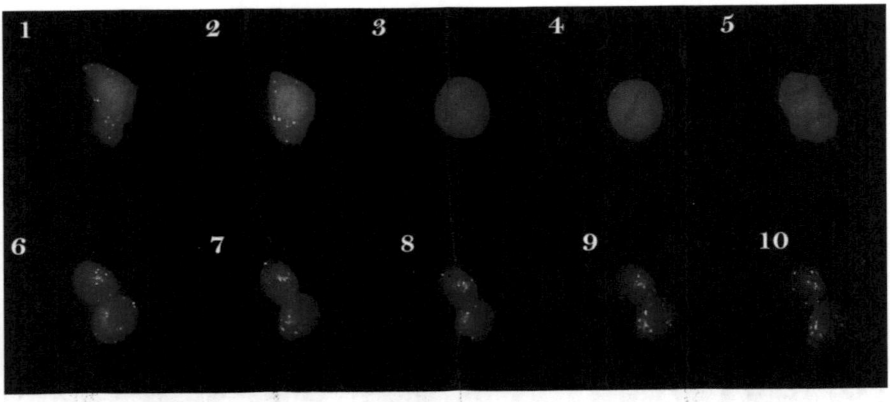


**Fig. 3.17** Autophagy is inhibited during mitosis in a single MCF7 cell

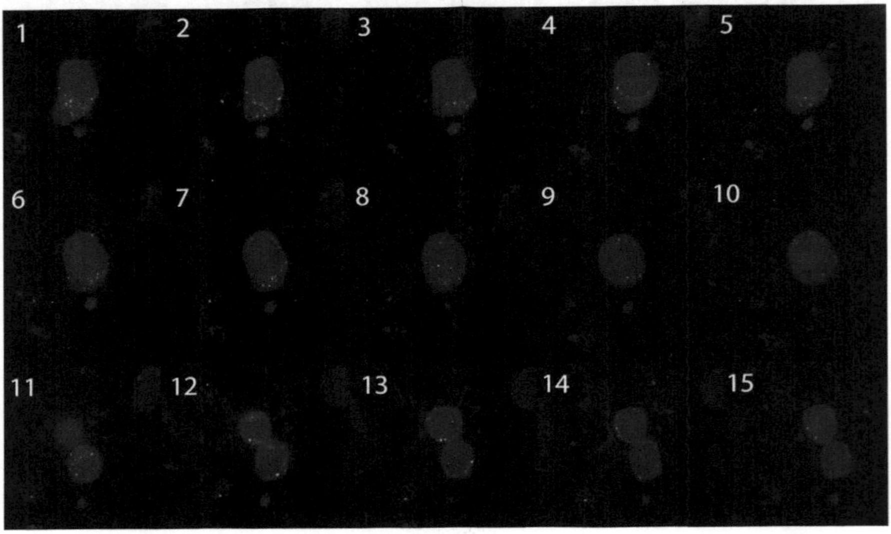


**Fig. 3.18** An MCF7 cell expressing GFP-LC3 during mitosis

These controversies can be due to lack of accurate quantitative methods to track autophagy (different stages of autophagy) during cell cycle phases. Development of such methods is necessary to study the dynamics of autophagy and requires using different markers for different stages of autophagy as a multistep process.

We have observed the mitosis of MCF-7 cells expressing GFP-LC3 and our results are based on looking at GFP-LC3 in these cells. Obviously GFP-LC3 puncta have disappeared during mitosis in the single MCF7 cell shown in Fig. 3.17 and movie 15. Time-lapse microscopy was done every hour in this case. But the other MCF7 cells shown in Figs. 3.18 and 3.19 (Movies 16 and 17) have kept their GFP-LC3 puncta during mitosis. Time-lapse microscopy was done every 15 min

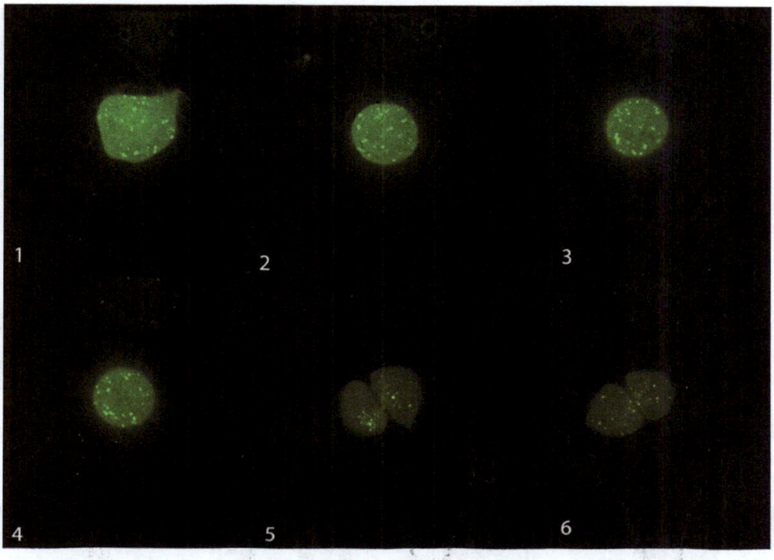


**Fig. 3.19** An MCF7 cell expressing GFP-LC3 during mitosis

in these cases. Although these are just some observations, they indicate importance of considering cell cycle phases and cell division when looking at autophagy as a dynamic response in time and space. Movies 18.1, 18.2, and 18.3 also show time-lapse microscopy of a single H4 cell going through cell division (Movie 18.1 shows GFP-LC3, Movie 18.2 shows H2B-RFP, and Movie 18.3 shows the merged version of Movies 18.1 and 18.2). Time-lapse microscopy was done every 15 min for this cell and the H2B-RFP in the red channel shows how mitosis happens. This cell continues autophagy during mitosis as well.

### *3.4.3 Serum Starvation-Induced Autophagy in Breast Cancer Cells*

The growth factor signaling pathway crosstalks with the autophagy pathway. Serum withdrawal, one of the stressors to induce autophagy in cancer cells, also removes growth factor from them [25, 36]. Here we have used serum starvation as a stressor to induce autophagy and compare its dynamics in MCF7, LCC1, and LCC9 cells (Fig. 3.20).

We tracked temporal changes in autophagy level (marked by GFP-LC3 puncta area) in single MCF7, LCC1, and LCC9 cells. Figure 3.21 and movie 19 are time-lapse microscopy of a single MCF7 cell expressing GFP-LC3 under serum starvation condition, and Fig. 3.22 shows the quantification of temporal changes in GFP-LC3 puncta area fold change in this cell (images were taken every 1 h).

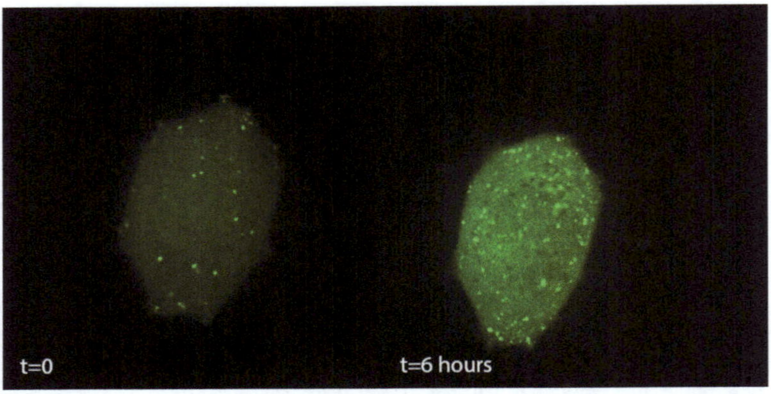


**Fig. 3.20** Serum starvation induces autophagic response in a single LCC9 cell expressing GFP-LC3

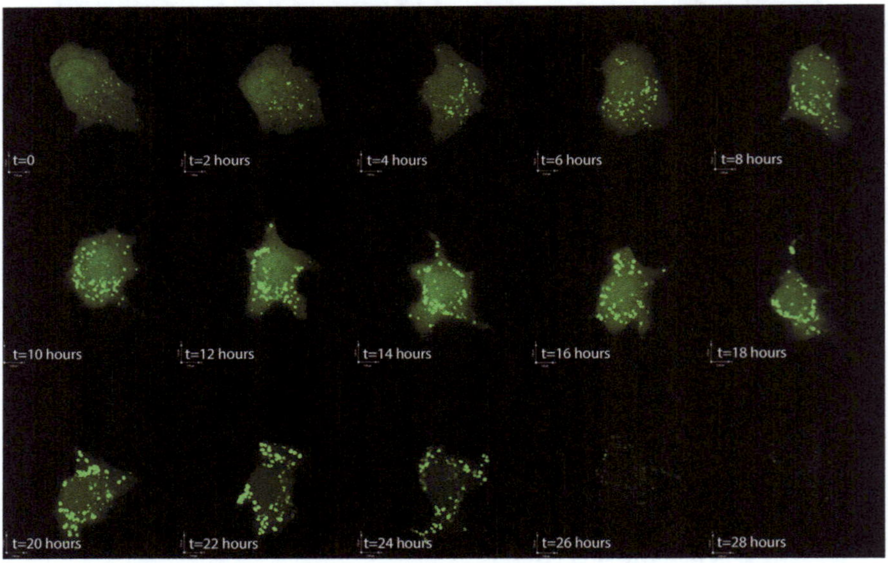


**Fig. 3.21** Time-lapse microscopy of an MCF7 cell expressing GFP-LC3 under serum starvation condition

This MCF7 cell increased its autophagy level and could survive even after 24 h of serum starvation. The GFP-LC3 signal was disappeared after 28 h which might be due to cell death. This pattern was seen in other single MCF7 cells as well after serum withdrawal (Movie 20, images were taken every 1 h). By serum starvation, we can control the duration of the stress, and try to find more quantitative information for getting a stress/response curve such as what the mathematical model in Chap. 2 predicts (Fig. 2.4). For example in one case, we removed and added the serum back

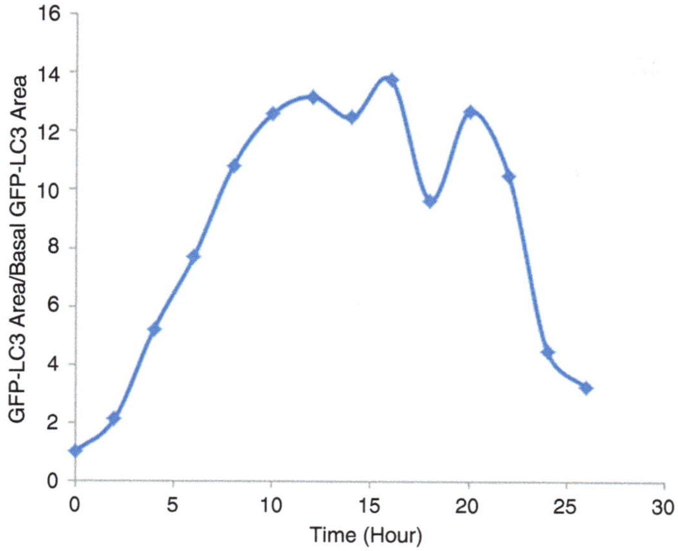


**Fig. 3.22** Time course of GFP-LC3 puncta area fold change in a single MCF7 cell under serum starvation condition

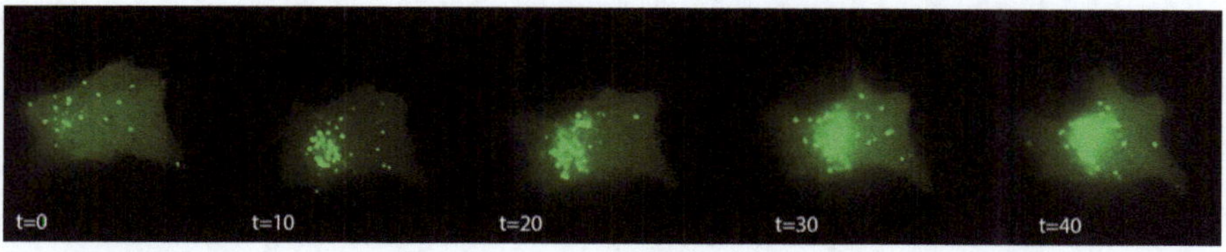


**Fig. 3.23** Time-lapse microscopy of autophagy characterized by GFP-LC3 in a single MCF7 cell under 10-h serum starvation condition

after 10 h. In this experiment, the single MCF7 cell could survive by increasing its autophagy level (Fig. 3.23 and Movie 21, images were taken every 15 min).

Quantification of GFP-LC3 puncta area fold change in this case revealed the same dynamics as the mathematical model describes: autophagy level increases and cell survives with a higher steady-state level of autophagy (Figs. 2.4 and 3.24).

In LCC1 and LCC9 cells, based on what we observed in some single cells, the autophagic response is more acute than MCF7 cells. The main difference between MCF7 cells and LCC1 and LCC9 cells is estrogen dependency. More investigations are needed to connect estrogen receptor signaling, growth factor signaling and autophagy.

For example, the LCC1 cell shown in Fig. 3.25 increases its autophagy after serum starvation, and apoptotic morphology appears after 135 min (Movie 22).

Quantification of GFP-LC3 puncta area fold change shows the same pattern as what we had observed in MCF7 cells, but with more acute kinetics (Fig. 3.26).

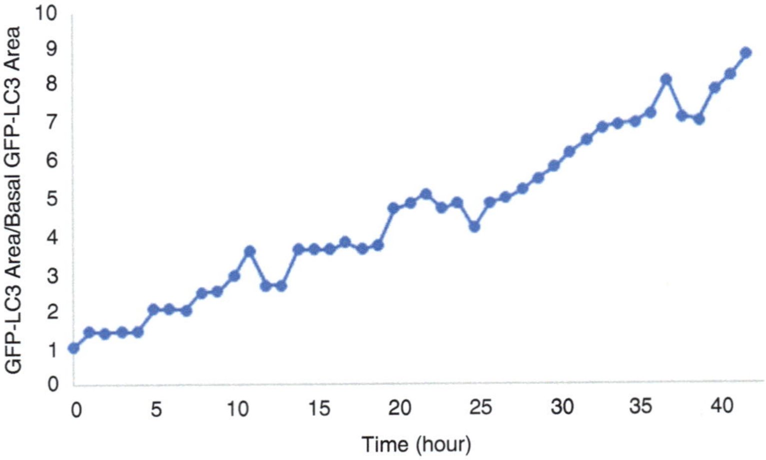


**Fig. 3.24** Time course of autophagy characterized by GFP-LC3 area fold change in a single MCF7 cell under 10-h serum starvation condition

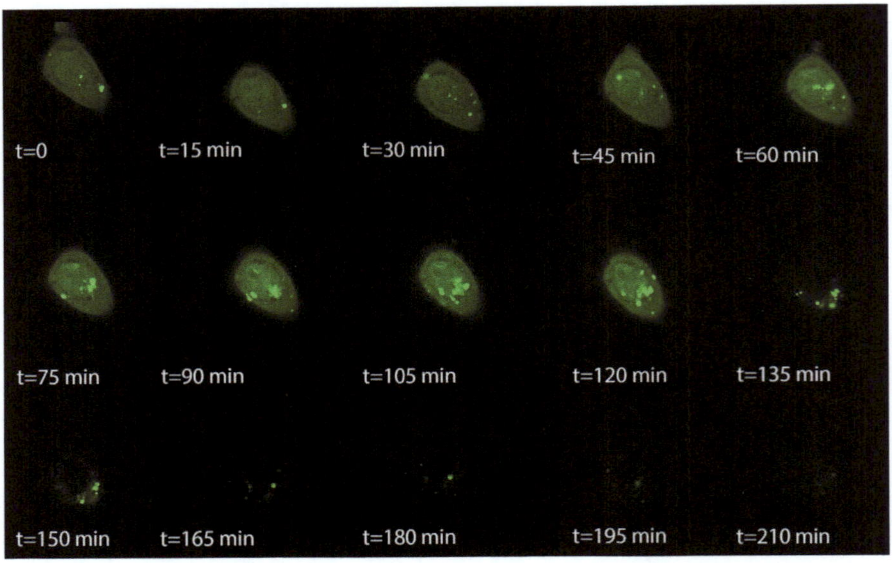


**Fig. 3.25** Time-lapse microscopy of an LCC1 cell expressing GFP-LC3 under serum starvation condition

In single LCC9 cells also, serum starvation-induced autophagy was a more acute response compared to MCF7 cells. Figures 3.27 (Movie 23, images were taken every 1 h) and 3.28 (Movie 24, images were taken every 15 min) show how serum-starved single LCC9 cells reach apoptotic morphology in 8 h and 105 min,

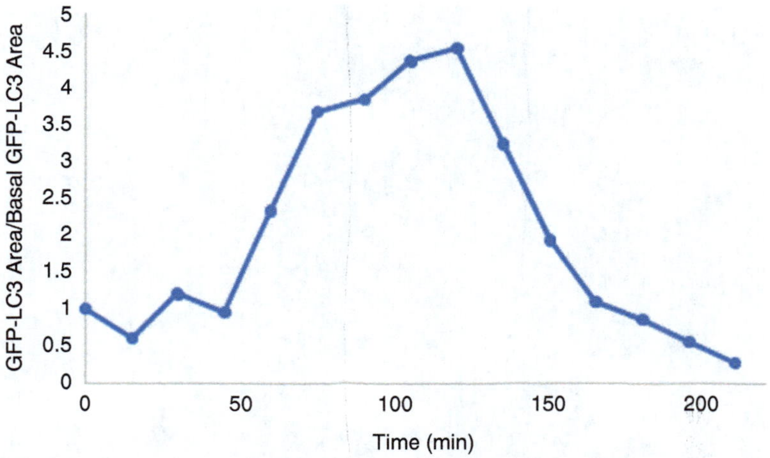


**Fig. 3.26** Time course of GFP-LC3 puncta area fold change in a single LCC1 cell under serum starvation condition

after increasing their autophagy level. Quantification of GFP-LC3 puncta area fold change is presented in Figs. 3.29 and 3.30 respectively.

Modulating the stress time (decreasing the duration of serum starvation) is also a good tool to capture the relationship between stress duration and autophagic response. Figures 3.31 and 3.32 show how a single LCC9 cell responds to 3-h serum starvation by increasing its autophagy level and keeping its survival for 9 h (Movie 25, images were taken every 1 h).

In this experiment, after removal, serum was given back to the cell after 3 h.

## 3.5 Current Realities and Future Directions

The large variability in cellular stress responses in different single cells within the heterogenous cell populations leads to a distribution of cell fate and responses. For example, in case of processes which functions as bistable switches, bimodality emerges at population level [2, 5, 16, 50]. To understand the systems level temporal behavior of cells, quantifying the time course and dynamical features in single live cells is an important step [42].

One of the most important questions in the field of autophagy is the existence of "Autophagic Cell Death" as a mode of cell death, which is different from apoptosis and necrosis [31, 37]. Autophagic cell death is defined in morphological terms: cytoplasmic accumulation of autophagosomes (marked by LC3) and simultaneous cell death. Whether this cell death is "cell death by autophagy" or "cell death with autophagy" is not clear. Answering this question can shed light on new directions for decoding the temporal behaviors of autophagy. Based on our observations,

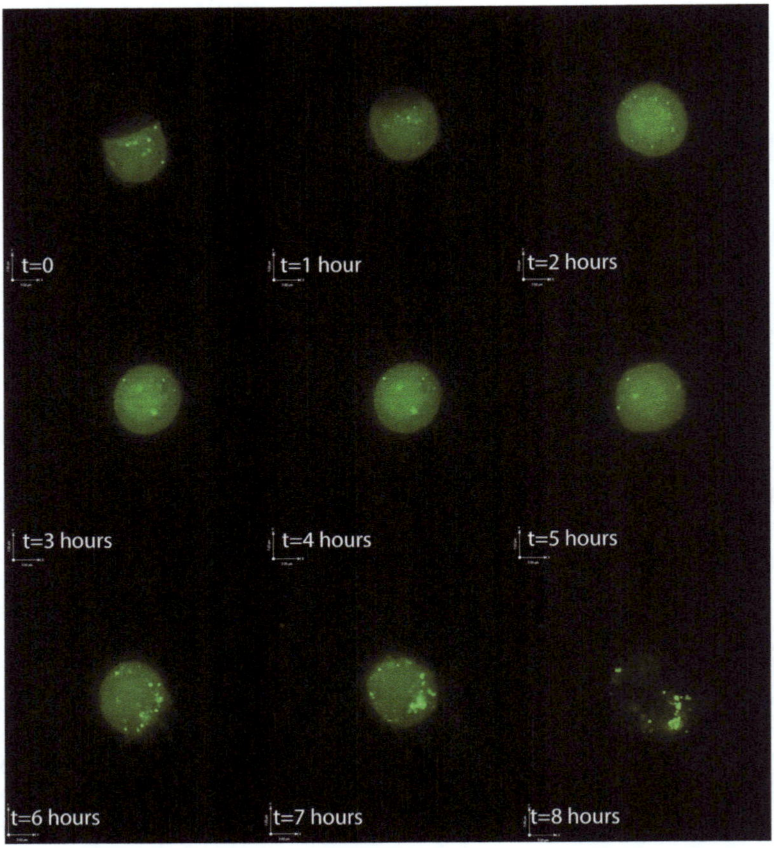


**Fig. 3.27** Time-lapse microscopy of an LCC9 cell expressing GFP-LC3 under serum starvation condition

morphologically, there are two scenarios for cell death after induction of autophagy. In H4 neuroglioma cells cisplatin treatment increases autophagy, and cell death may occur if cisplatin dose is large enough. Some cells increase the basal autophagy (GFP-LC3 puncta area), and after reaching the maximum autophagy flux, GFP-LC3 level decreases to a very low level, so that there will not be morphologically significant GFP-LC3 puncta in cytoplasm. Cell death happens after this morphology is formed (H2B-RFP degradation). Upper panel in Fig. 3.33 shows this scenario in a single H4 neuroglioma cell. It seems that this scenario represents apoptosis after autophagic response.

Some H4 neuroglioma cells increase their basal autophagy after cisplatin treatment, and after reaching maximum autophagy flux (GFP-LC3 puncta area), cell death occurs (represented by H2B-RFP degradation). In this case, the cell has degradation of H2B-RFP combined with accumulation of GFP-LC3 in cytoplasm. This morphology can be what is referred to as "Autophagic Cell Death" in the

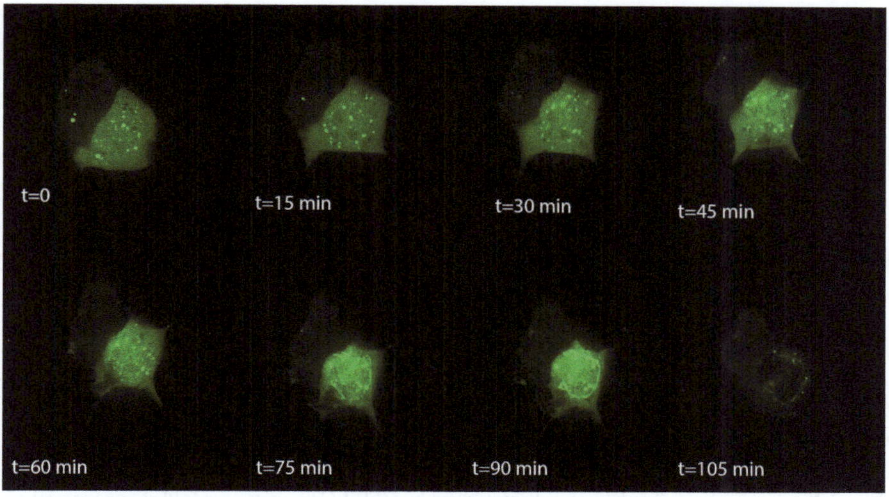


**Fig. 3.28** Time-lapse microscopy of autophagy characterized by GFP-LC3 in a single LCC9 cell under serum starvation condition

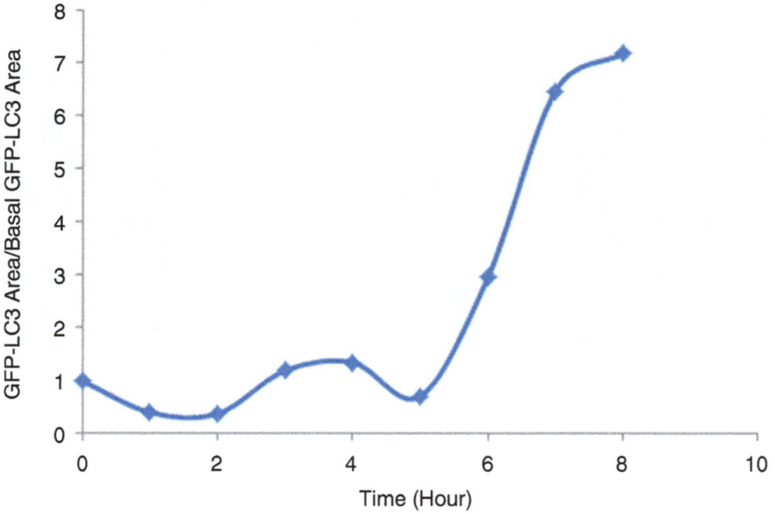


**Fig. 3.29** Time course of GFP-LC3 puncta area fold change in a single LCC9 cell under serum starvation condition

literature. The lower panel in Fig. 3.33 shows this scenario in a H4 neuroglioma cell treated with cisplatin. How each of these scenarios happens seems to have an answer related to the dynamic relationship of autophagy and apoptosis. Measuring the time courses of autophagic flux and initiation of cell death in a large sample of single cells would be a useful method to see under what conditions these scenarios can take place.

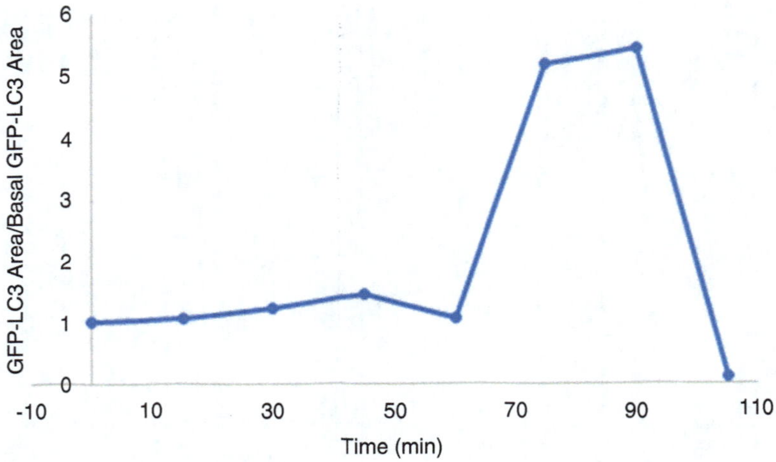


**Fig. 3.30** Time course of autophagy characterized by GFP-LC3 puncta area fold change in a single LCC9 cell under serum starvation condition

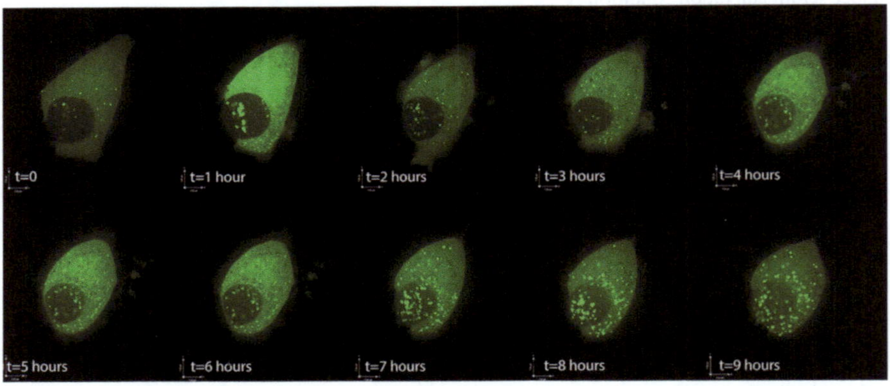


**Fig. 3.31** Time-lapse microscopy of an LCC9 cell expressing GFP-LC3 under 3-h serum starvation condition

The framework and examples presented in this chapter can be the initial steps to quantify autophagy as a dynamic response in therapeutic-resistant cancer cells, such as LCC9 cells, at single-cell level. Because autophagy takes part in evolution of endocrine-resistance, comparing the kinetic parameters of autophagic response in resistant and sensitive cells is a key to reveal dynamical determinants of endocrine resistance in breast cancer.

Autophagy may persist chronically in some cancer cells dealing with special stress condition (MCF7 cells and serum starvation); hence, we need to use methods to be able to capture the dynamical features of this response with regard to mitosis and cell cycle phases, because cell cycle phases affect the autophagy level [17, 18, 43, 49]. Development of novel quantitative imaging methods based on fixed

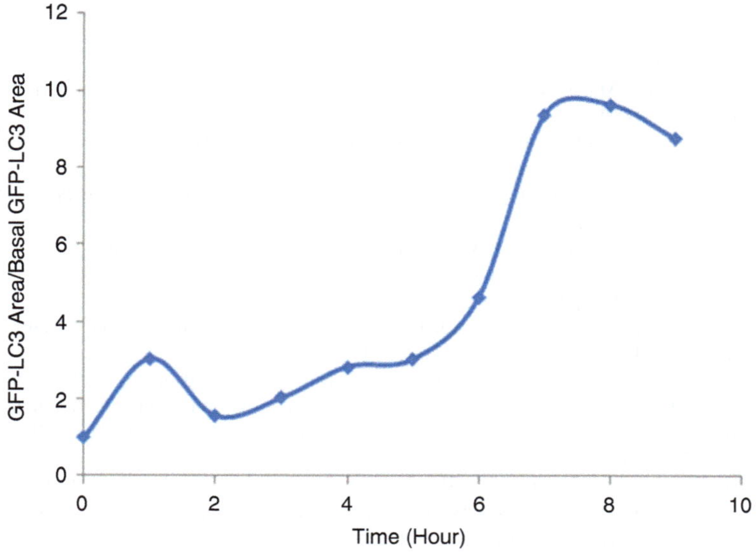


**Fig. 3.32** Time course of GFP-LC3 puncta area fold change in a single LCC9 under serum starvation condition

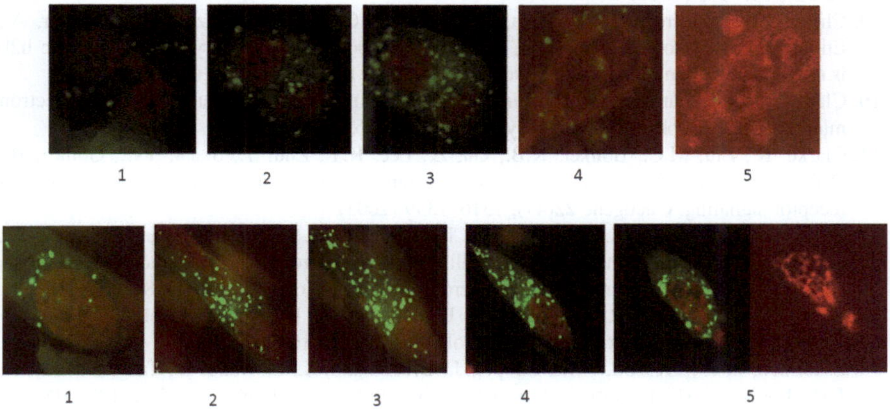


**Fig. 3.33** Morphology of cell death due to autophagy

cells can be used to overcome this issue [26]. The future experimental efforts should be on high-throughput quantitative analysis and live cell imaging of the interplay of autophagy and apoptosis in single cells in 3D space. Application of markers for targets of autophagy such as ER tracker or mitochondrial tracker is helpful in decoding dynamics of selective autophagy.

## References


1. Abràmoff, M.D., Magalhães, P.J., Ram, S.J.: Image processing with imagej. Biophoton. Int. **11**(7), 36–42 (2004)
2. Alon, U.: Network motifs: theory and experimental approaches. Nat. Rev. Genet. **8**(6), 450–461 (2007)
3. Ashford, T.P., Porter, K.R.: Cytoplasmic components in hepatic cell lysosomes. J. Cell Biol. **12**(1), 198–202 (1962)
4. Bampton, E.T.W., Goemans, C.G., Niranjan, D., Mizushima, N., Tolkovsky, A.M.: The dynamics of autophagy visualised in live cells: from autophagosome formation to fusion with endo/lysosomes. Autophagy **1**(1), 23–36 (2005)
5. Biggar, S.R., Crabtree, G.R.: Cell signaling can direct either binary or graded transcriptional responses. EMBO J. **20**(12), 3167–3176 (2001)
6. Brünner, N., Frandsen, T.L., Holst-Hansen, C., Bei, M., Thompson, E.W., Wakeling, A.E., Lippman, M.E., Clarke, R.: MCF7/LCC2: a 4-hydroxytamoxifen resistant human breast cancer variant that retains sensitivity to the steroidal antiestrogen ICI 182,780. Cancer Res. **53**(14), 3229–3232 (1993)
7. Brünner, N., Boysen, B., Jirus, S., Skaar, T.C., Holst-Hansen, C., Lippman, J., Frandsen, T., Spang-Thomsen, M., Fuqua, S.A.W., Clarke, R.: Mcf7/lcc9: an antiestrogen-resistant mcf-7 variant in which acquired resistance to the steroidal antiestrogen ici 182,780 confers an early cross-resistance to the nonsteroidal antiestrogen tamoxifen. Cancer Res. **57**(16), 3486–3493 (1997)
8. Chen, Y., Klionsky, D.J.: The regulation of autophagy–unanswered questions. J. Cell Sci. **124**(2), 161–170 (2011)
9. Cheung, W.L., Ajiro, K., Samejima, K., Kloc, M., Cheung, P., Mizzen, C.A., Beeser, A., Etkin, L.D., Chernoff, J., Earnshaw, W.C., et al.: Apoptotic phosphorylation of histone h2b is mediated by mammalian sterile twenty kinase. Cell **113**(4), 507–517 (2003)
10. Clark, S.L.: Cellular differentiation in the kidneys of newborn mice studied with the electron microscope. J. Biophys. Biochem. Cytol. **3**(3), 349–362 (1957)
11. Clarke, R., Liu, M.C., Bouker, K.B., Gu, Z., Lee, R.Y., Zhu, Y., Skaar, T.C., Gomez, B., O'Brien, K., Wang, Y., et al.: Antiestrogen resistance in breast cancer and the role of estrogen receptor signaling. Oncogene **22**(47), 7316–7339 (2003)
12. Clarke, R., Shajahan, A.N., Riggins, R.B., Cho, Y., Crawford, A., Xuan, J., Wang, Y., Zwart, A., Nehra, R., Liu, M.C.: Gene network signaling in hormone responsiveness modifies apoptosis and autophagy in breast cancer cells. J. Steroid Biochem. Mol. Biol. **114**(1), 8–20 (2009)
13. Deter, R.L., Baudhuin, P., De Duve, C.: Participation of lysosomes in cellular autophagy induced in rat liver by glucagon. J. Cell Biol. **35**(2), C11–C16 (1967)
14. Eskelinen, E.-L., Prescott, A.R., Cooper, J., Brachmann, S.M., Wang, L., Tang, X., Backer, J.M., Lucocq, J.M.: Inhibition of autophagy in mitotic animal cells. Traffic **3**(12), 878–893 (2002)
15. Eskelinen, E.-L., Reggiori, F., Baba, M., Kovács, A.L., Seglen, P.O.: Seeing is believing: the impact of electron microscopy on autophagy research. Autophagy **7**(9), 935–956 (2011)
16. Ferrell, J.E., Machleder, E.M.: The biochemical basis of an all-or-none cell fate switch in xenopus oocytes. Science **280**(5365), 895–898 (1998)
17. Filippi-Chiela, E.C., Villodre, E.S., Zamin, L.L., Lenz, G.: Autophagy interplay with apoptosis and cell cycle regulation in the growth inhibiting effect of resveratrol in glioma cells. PLoS One **6**(6), e20849 (2011)
18. Furuya, T., Kim, M., Lipinski, M., Li, J., Kim, D., Lu, T., Shen, Y., Rameh, L., Yankner, B., Tsai, L.-H., et al.: Negative regulation of VPS34 by CDK mediated phosphorylation. Mol. Cell **38**(4), 500–511 (2010)
19. Goldstein, J.C., Kluck, R.M., Green, D.R.: A single cell analysis of apoptosis: ordering the apoptotic phenotype. Ann. N. Y. Acad. Sci. **926**(1), 132–141 (2000)

20. Goldstein, J.C., Waterhouse, N.J., Juin, P., Evan, G.I., Green, D.R.: The coordinate release of cytochrome c during apoptosis is rapid, complete and kinetically invariant. Nat. Cell Biol. **2**(3), 156–162 (2000)
21. Goldstein, J.C., Munoz-Pinedo, C., Ricci, J.E., Adams, S.R., Kelekar, A., Schuler, M., Tsien, R.Y., Green, D.R.: Cytochrome c is released in a single step during apoptosis. Cell Death Differ. **12**(5), 453–462 (2005)
22. Haley, E.M.: Autophagy induction in the quiescent state: in vitro and in vivo characterization of cellular self-cannibalism in quiescent cells. Ph.D. thesis, Princeton University (2011)
23. Harhaji-Trajkovic, L., Vilimanovich, U., Kravic-Stevovic, T., Bumbasirevic, V., Trajkovic, V.: AMPK-mediated autophagy inhibits apoptosis in cisplatin-treated tumour cells. J. Cell. Mol. Med. **13**(9b), 3644–3654 (2009)
24. Heatwole, V.M.: Tunel assay for apoptotic cells. In: Immunocytochemical Methods and Protocols, pp. 141–148. Springer, New York (1999)
25. Janji, B., Viry, E., Baginska, J., Van Moer, K., Berchem, G.: Role of autophagy in cancer and tumor progression. Autophagy: A double-edged sword-Cell survival or death, 189–215 (2013)
26. Kafri, R., Levy, J., Ginzberg, M.B., Oh, S., Lahav, G., Kirschner, M.W.: Dynamics extracted from fixed cells reveal feedback linking cell growth to cell cycle. Nature **494**(7438), 480–483 (2013)
27. Kaminskyy, V., Abdi, A., Zhivotovsky, B.: A quantitative assay for the monitoring of autophagosome accumulation in different phases of the cell cycle. Autophagy **7**(1), 83–90 (2011)
28. Katayama, H., Kogure, T., Mizushima, N., Yoshimori, T., Miyawaki, A.: A sensitive and quantitative technique for detecting autophagic events based on lysosomal delivery. Chem. Biol. **18**(8), 1042–1052 (2011)
29. Klionsky, D.J., Cuervo, A.M., Seglen, P.O.: Methods for monitoring autophagy from yeast to human. Autophagy **3**(3), 181–206 (2007)
30. Klionsky, D.J., Abdalla, F.C., Abeliovich, H., Abraham, R.T., Acevedo-Arozena, A., Adeli, K., Agholme, L., Agnello, M., Agostinis, P., Aguirre-Ghiso, J.A., et al.: Guidelines for the use and interpretation of assays for monitoring autophagy. Autophagy **8**(4), 445–544 (2012)
31. Kroemer, G., Levine, B.: Autophagic cell death: the story of a misnomer. Nat. Rev. Mol. Cell Biol. **9**(12), 1004–1010 (2008)
32. Liang, C.: Negative regulation of autophagy. Cell Death Differ. **17**(12), 1807–1815 (2010)
33. Liu, L., Xie, R., Nguyen, S., Ye, M., McKeehan, W.L.: Robust autophagy/mitophagy persists during mitosis. Cell Cycle **8**(10), 1616–1620 (2009)
34. Loewer, A., Lahav, G.: We are all individuals: causes and consequences of non-genetic heterogeneity in mammalian cells. Curr. Opin. Genet. Dev. **21**(6), 753–758 (2011)
35. Loos, B., Engelbrecht, A.-M.: Cell death: a dynamic response concept. Autophagy **5**(5), 590–603 (2009)
36. Lum, J.J., Bauer, D.E., Kong, M., Harris, M.H., Li, C., Lindsten, T., Thompson, C.B.: Growth factor regulation of autophagy and cell survival in the absence of apoptosis. Cell **120**(2), 237–248 (2005)
37. Maiuri, M.C., Zalckvar, E., Kimchi, A., Kroemer, G.: Self-eating and self-killing: crosstalk between autophagy and apoptosis. Nat. Rev. Mol. Cell Biol. **8**(9), 741–752 (2007)
38. Mizushima, N., Levine, B., Cuervo, A.M., Klionsky, D.J.: Autophagy fights disease through cellular self-digestion. Nature **451**(7182), 1069–1075 (2008)
39. Ogata, M., Hino, S.-I., Saito, A., Morikawa, K., Kondo, S., Kanemoto, S., Murakami, T., Taniguchi, M., Tanii, I., Yoshinaga, K., et al.: Autophagy is activated for cell survival after endoplasmic reticulum stress. Mol. Cell. Biol. **26**(24), 9220–9231 (2006)
40. Pankiv, S., Clausen, T.H., Lamark, T., Brech, A., Bruun, J.-A., Outzen, H., Øvervatn, A., Bjørkøy, G., Johansen, T.: P62/SQSTM1 binds directly to ATG8/LC3 to facilitate degradation of ubiquitinated protein aggregates by autophagy. J. Biol. Chem. **282**(33), 24131–24145 (2007)
41. Proikas-Cezanne, T., Codogno, P.: A new fluorescence-based assay for autophagy. Chem. Biol. **18**(8), 940–941 (2011)

42. Purvis, J.E., Lahav, G.: Encoding and decoding cellular information through signaling dynamics. Cell **152**(5), 945–956 (2013)
43. Rubinsztein, D.C.: CDKS regulate autophagy via VPS34. Mol. Cell **38**(4), 483–484 (2010)
44. Saraste, A.: Morphologic criteria and detection of apoptosis. Herz **24**(3), 189–195 (1999)
45. Sgonc, R., Wick, G.: Methods for the detection of apoptosis. Int. Arch. Allergy Immunol. **105**(4), 327–332 (2009)
46. Shvets, E., Fass, E., Elazar, Z.: Utilizing flow cytometry to monitor autophagy in living mammalian cells. Autophagy **4**(5), 621–628 (2008)
47. Spencer, S.L., Sorger, P.K.: Measuring and modeling apoptosis in single cells. Cell **144**(6), 926–939 (2011)
48. Tanida, I., Minematsu-Ikeguchi, N., Ueno, T., Kominami, E.: Lysosomal turnover, but not a cellular level, of endogenous LC3 is a marker for autophagy. Autophagy **1**(2), 84–91 (2005)
49. Tasdemir, E., Maiuri, M.C., Tajeddine, N., Vitale, I., Criollo, A., Vicencio, J.M., Hickman, J.A., Geneste, O., Kroemer, G.: Cell cycle-dependent induction of autophagy, mitophagy and reticulophagy. Cell Cycle **6**(18), 2263–2267 (2007)
50. Tyson, J.J., Baumann, W.T., Chen, C., Verdugo, A., Tavassoly, I., Wang, Y., Weiner, L.M., Clarke, R.: Dynamic modelling of oestrogen signalling and cell fate in breast cancer cells. Nat. Rev. Cancer **11**(7), 523–532 (2011)
51. Vermes, I., Haanen, C., Steffens-Nakken, H., Reutellingsperger, C.: A novel assay for apoptosis flow cytometric detection of phosphatidylserine expression on early apoptotic cells using fluorescein labelled annexin v. J. Immunol. Methods **184**(1), 39–51 (1995)
52. Wang, R.C., Levine, B.: Autophagy in cellular growth control. FEBS Lett. **584**(7), 1417–1426 (2010)
53. Wells, W.A.: Autophagy unveiled. J. Cell Biol. **168**(5), 676–677 (2005)
54. Wrighton, K.H.: Autophagy: kinase ups and downs. Nat. Rev. Mol. Cell Biol. **11**(7), 464 (2010)
55. Xu, Y., Yuan, J., Lipinski, M.M.: Live imaging and single-cell analysis reveal differential dynamics of autophagy and apoptosis. Autophagy **9**(9), 0–1 (2013)
56. Yao, K.C., Komata, T., Kondo, Y., Kanzawa, T., Kondo, S., Germano, I.M.: Molecular response of human glioblastoma multiforme cells to ionizing radiation: cell cycle arrest, modulation of cyclin-dependent kinase inhibitors, and autophagy. J. Neurosurg. **98**(2), 378–384 (2003)

# Chapter 4
# Conclusions: Future Directions in Systems Biology of Autophagy

Mathematical models of physiological processes in cells, tissues, and organs can give integrative insights on how homeostasis is controlled under physiological conditions at a systems level. Mathematical models can also show how physiologic functions turn into pathophysiological processes. These mathematical models can be applied in combination with experimental methods to explain and predict the behaviors of biological systems [1]. When combined with quantitative and qualitative experimental methodologies, these models can be used to guide future experiments [26].

Some examples of how mathematical models have been used in understanding complex diseases include models for population dynamics of red blood cells in anemia [18], breast cancer anti-estrogen sensitivity and resistance [26, 34], combination therapy response in cancer [2, 6], and the immune system responses in health and disease [14, 19].

Designing *in silico* models of signaling networks of cellular functions may aid in drug discovery and in finding optimized therapeutic regimens [3, 10, 11, 20, 35, 41]. In the context of systems pharmacology, the goal is to discover connections among signaling networks, their dynamics, and the final physiological functions that are perturbed in diseases such as cancer [3, 7, 30, 32, 33, 35, 41]. Therapeutic resistance is one of the major challenges in oncology practice and cancer biology [12, 16]. Because the mechanisms leading to drug resistance are very complex (including interplay of many cellular processes, such as autophagy, UPR, and apoptosis), quantitative analysis of the signaling networks can help to find novel strategies to overcome the therapeutic-resistant phenotypes in tumors [26, 34].

We presented a basic mathematical model of the interaction of autophagy and apoptosis in Chap. 2 and defined an experimental methodology for validation of the mathematical model in Chap. 3. The model predictions are consistent with quantitative data available in literature [27, 40], and some of the data presented

 
I. Tavassoly, *Dynamics of Cell Fate Decision Mediated by the Interplay of Autophagy and Apoptosis in Cancer Cells*, Springer Theses, DOI 10.1007/978-3-319-14962-2_4

in Chap. 3 shows the same dynamics predicted by the model. A very good example of validation of the model presented in Chap. 2 is the results from Xu et al. [40] on time course of autophagic response and cell fate.

Xu et al. [40] have presented some aspects of the dynamics predicted by the mathematical model described in Chap. 2. The first version of this mathematical model which carried just basic components of signaling network of autophagy and apoptosis in breast cancer cells was presented in 2011 [34], when quantitative data were not available to validate the model predictions. But Xu et al. [40] have taken a new road to the quantitative study of autophagy and cell fate mediated by it. The mathematical model presented in Chap. 2, while was developed before the quantitative data became available, described autophagy to function like a rheostat compared to the on/off switch (bistable switch) of dynamics of apoptosis.

In the mathematical model autophagy is induced first to delay and suppress apoptosis [34]. Xu et al. [40] have found that autophagy induction in response to different forms of stress is uniformly unimodal compared to the bimodal and all-or-none dynamics of apoptosis. They have hypothesized that autophagy is a process which is graded in nature and dynamically is adjusted to keep the homeostasis of an individual cell or a cell population under stress [40]. They have shown that starvation and mTOR inhibitors trigger autophagy in a less acute fashion compared to autophagy induction due to cytotoxic agents. It is somehow physiologically important for cells to have a more robust and acute autophagic response when the stress is strong and acute and can threaten the cell survival. From the viewpoint of the mathematical model presented in Chap. 2, induction of autophagy through mTOR inhibition (starvation, rapamycin treatment) is less effective and less acute compared to autophagy induced by cytotoxic or endocrine therapies, because inhibition of mTOR contributes to autophagosome formation via just one part of the molecular circuit presented in Figs. 1.2, 1.5, and 2.3, while cytotoxic agents use the whole network to induce autophagy.

It has been evident that autophagy is a cytoprotective process which postpones apoptosis. Xu et al. [40] have proved that autophagy induced by cytotoxic agent (STS) delays apoptosis but cannot fully prevent it. This fact is consistent with the model predictions presented in Fig. 2.4 indicating autophagy works like a resistor or rheostat.

Xu et al. [40] have found that very strong stress, such as high dose of STS, long exposure to STS, or long-term glucose starvation, can decrease autophagy level. This phenomenon is consistent with the dynamics described by the mathematical model presented in Chap. 2. When stress is strong, apoptosis is initiated sooner, and, due to caspase activities and cleavage of Beclin-1, autophagy level decreases. Panel A in Fig. 2.4 shows that when RPTC cells are treated with higher doses of cisplatin (such as 24 $\mu$M cisplatin), autophagy level comes down faster and autophagic flux will not reach the level achieved by lower doses of cisplatin (such as 18 $\mu$M). It has been shown that UPR is induced before apoptosis and has a cytoprotective role. Figure 2.2 summarizes the interaction between UPR, autophagy, and apoptosis. As a matter of fact, UPR transduces stress signal to autophagy module if it is not successful to suppress the stress [4, 5, 21, 22, 34]. Xu et al. [40] have observed that tunicamycin, which triggers UPR, induces autophagic response with a slower dynamics. From the modeling hypothesis presented in Chap. 2, UPR itself can inhibit the stress and contribute to cell survival. This observation by Xu et al. [40] shows that if the stress signal exclusively passes through the UPR module (tunicamycin treatment), due to prosurvival role of UPR, the signal is attenuated by the time it is received by the autophagy module.

In general, the dynamic model presented in Chap. 2 can consistently predict the quantitative data provided by Xu et al. [40]. The experimental framework presented in Chap. 3 can be used to study the dynamics of autophagic response in single cancer cells including therapeutic resistant cancer cells such as LCC9 and LCC2 cells. Some of the time course data provided by this framework is consistent with predictions of the mathematical model, and more single-cell data using this experimental setting can help to validate the model and find stress/response curves and accurate parameter sets for ODEs in different types of cells.

Another step, beyond quantitative imaging of cell lines, is quantitative and live-cell imaging of signaling pathways and their dynamics in real tumors, which will provide more realistic data on cancer development and progression. *In vivo* imaging of tumor development in animal models and patients gives us realistic data for understanding its dynamics. Biomolecular imaging techniques such as positron emission tomography (PET) [15], Magnetic Resonance Imaging (MRI), and Fluorescence Molecular Tomography (FMT) have been used to track biological processes like cell death and metastasis in cancer [37, 38]. Recently, imaging methods to visualize cellular interactions *in vivo* have been introduced as well [29]. Using novel technologies to quantify the stress responses such as autophagy in tumors of patients or animal models can give a better understanding of the interaction of autophagy and apoptosis. Fluorescence Tomography has been used to image autophagy in the heart *in vivo* [8] and a GFP-LC3 transgenic mouse has been used to detect autophagy in a stroke model *in vivo* [31].

Another step beyond the current mathematical model will be adding signaling networks of UPR, necrosis, p53 pathway, growth factor signaling, cell cycle, cell growth, senescence, quiescence, etc. to the current wiring diagram of the model.

Tumor responses to treatment can be best described mathematically by multiscale modeling. Dynamics of signaling networks in single cells can be connected to cell population dynamics, contributing to tumor growth or shrinkage using PDEs. Effects of tumor microenvironment, such as the reverse Warburg effect, can be modeled at a multiscale level. In a multiscale model of tumor response to therapeutics, cancer stem cells must be considered, because they participate in development of therapeutic resistance in tumors [9, 17, 28]. On the other hand, stem cells, including cancer stem cells, have their own mechanisms of cell fate control, and the autophagic response in stem cells contributes to their "Stemness" [23–25]. Autophagy in stem cells is a mechanism of development of chemoresistance in cancer [39].

MicroRNAs are an important part of signaling networks controlling cell fate in cancer cells [36]. For a virtual model of cancer cell, we need to add microRNAs to mathematical models. For example, autophagic response is controlled by microRNAs, and it is necessary to add microRNAs involved in control of autophagy to the current version of the mathematical model of the interplay of autophagy and apoptosis [13, 42].

## References


1. Alon, U.: Introduction to Systems Biology: And the Design Principles of Biological Networks, vol. 10. CRC Press, Boca Raton (2007)
2. Bagheri, N., Shiina, M., Lauffenburger, D.A., Korn, W.M.: A dynamical systems model for combinatorial cancer therapy enhances oncolytic adenovirus efficacy by mek-inhibition. PLoS Comput. Biol. **7**(2), e1001085 (2011)
3. Berger, S.I., Iyengar, R.: Network analyses in systems pharmacology. Bioinformatics **25**(19), 2466–2472 (2009)
4. Bernales, S., McDonald, K.L., Walter, P.: Autophagy counterbalances endoplasmic reticulum expansion during the unfolded protein response. PLoS Biol. **4**(12), e423 (2006)
5. Bernales, S., Schuck, S., Walter, P.: Er-phagy. Autophagy **3**(3), 285–287 (2007)
6. Bozic, I., Reiter, J.G., Allen, B., Antal, T., Chatterjee, K., Shah, P., Moon, Y.S., Yaqubie, A., Kelly, N., Le, D.T., et al.: Evolutionary dynamics of cancer in response to targeted combination therapy. eLife **2**, e00747 (2013)
7. Butcher, E.C., Berg, E.L., Kunkel, E.J.: Systems biology in drug discovery. Nat. Biotechnol. **22**(10), 1253–1259 (2004)
8. Chen, H.H., Mekkaoui, C., Cho, H., Ngoy, S., Marinelli, B., Waterman, P., Nahrendorf, M., Liao, R., Josephson, L., Sosnovik, D.E.: Fluorescence tomography of rapamycin-induced autophagy and cardioprotection in vivo. Circ. Cardiovasc. Imaging **6**(3), 441–447 (2013)
9. Dean, M., Fojo, T., Bates, S.: Tumour stem cells and drug resistance. Nat. Rev. Cancer **5**(4), 275–284 (2005)
10. Ergün, A., Lawrence, C.A., Kohanski, M.A., Brennan, T.A., Collins, J.J.: A network biology approach to prostate cancer. Mol. Syst. Biol. **3**(1), 82 (2007)
11. Faratian, D., Goltsov, A., Lebedeva, G., Sorokin, A., Moodie, S., Mullen, P., Kay, C., Um, I.H., Langdon, S., Goryanin, I., et al.: Systems biology reveals new strategies for personalizing cancer medicine and confirms the role of pten in resistance to trastuzumab. Cancer Res. **69**(16), 6713–6720 (2009)
12. Fisher, G.A., Sikic, B.I.: Drug resistance in clinical oncology and hematology. Introduction. Hematol. Oncol. Clin. North Am. **9**(2), xi (1995)
13. Frankel, L.B., Lund, A.H.: Microrna regulation of autophagy. Carcinogenesis **33**(11), 2018–2025 (2012)
14. Fu, Y., Glaros, T., Zhu, M., Wang, P., Wu, Z., Tyson, J.J., Li, L., Xing, J.: Network topologies and dynamics leading to endotoxin tolerance and priming in innate immune cells. PLoS Comput. Biol. **8**(5), e1002526 (2012)
15. Gambhir, S.S.: Molecular imaging of cancer with positron emission tomography. Nat. Rev. Cancer **2**(9), 683–693 (2002)
16. Gottesman, M.M.: Mechanisms of cancer drug resistance. Ann. Rev. Med. **53**(1), 615–627 (2002)
17. Hanahan, D., Weinberg, R.A.: Hallmarks of cancer: the next generation. Cell **144**(5), 646–674 (2011)
18. Higgins, J.M., Mahadevan, L.: Physiological and pathological population dynamics of circulating human red blood cells. Proc. Natl. Acad. Sci. **107**(47), 20587–20592 (2010)
19. Hong, T., Xing, J., Li, L., Tyson, J.J.: A mathematical model for the reciprocal differentiation of t helper 17 cells and induced regulatory t cells. PLoS Comput. Biol. **7**(7), e1002122 (2011)

20. Kreeger, P.K., Lauffenburger, D.A.: Cancer systems biology: a network modeling perspective. Carcinogenesis **31**(1), 2–8 (2010)
21. Lee, D.Y., Lee, J., Sugden, B.: The unfolded protein response and autophagy: herpesviruses rule! J. Virol. **83**(3), 1168–1172 (2009)
22. Matus, S., Lisbona, F., Torres, M., León, C., Thielen, P., Hetz, C.: The stress rheostat: an interplay between the unfolded protein response (upr) and autophagy in neurodegeneration. Curr. Mol. Med. **8**(3), 157–172 (2008)
23. Mizushima, N., Yamamoto, A., Hatano, M., Kobayashi, Y., Kabeya, Y., Suzuki, K., Tokuhisa, T., Ohsumi, Y., Yoshimori, T.: Dissection of autophagosome formation using apg5-deficient mouse embryonic stem cells. J. Cell Biol. **152**(4), 657–668 (2001)
24. Mortensen, M., Soilleux, E.J., Djordjevic, G., Tripp, R., Lutteropp, M., Sadighi-Akha, E., Stranks, A.J., Glanville, J., Knight, S., Jacobsen, E.K.W., et al.: The autophagy protein atg7 is essential for hematopoietic stem cell maintenance. J. Exp. Med. **208**(3), 455–467 (2011)
25. Pan, H., Cai, N., Li, M., Liu, G.H., Belmonte, J.C.I.: Autophagic control of cell stemness. EMBO Mol. Med. **5**(3), 327–331 (2013)
26. Parmar, J.H., Cook, K.L., Shajahan-Haq, A.N., Clarke, P.A.G., Tavassoly, I., Clarke, R., Tyson, J.J., Baumann, W.T.: Modelling the effect of grp78 on anti-oestrogen sensitivity and resistance in breast cancer. Interface Focus **3**(4), 20130012 (2013)
27. Periyasamy-Thandavan, S., Jiang, M., Wei, Q., Smith, R., Yin, X.M., Dong, Z.: Autophagy is cytoprotective during cisplatin injury of renal proximal tubular cells. Kidney Int. **74**(5), 631–640 (2008)
28. Rich, J.N.: Cancer stem cells in radiation resistance. Cancer Res. **67**(19), 8980–8984 (2007)
29. Sellmyer, M.A., Bronsart, L., Imoto, H., Contag, C.H., Wandless, T.J., Prescher, J.A.: Visualizing cellular interactions with a generalized proximity reporter. Proc. Natl. Acad. Sci. **110**(21), 8567–8572 (2013)
30. Sorger, P.K., Allerheiligen, S.R.B., Abernethy, D.R., Altman, R.B., Brouwer, K.L.R., Califano, A., D'Argenio, D.Z., Iyengar, R., Jusko, W.J., Lalonde, R., et al.: Quantitative and systems pharmacology in the post-genomic era: new approaches to discovering drugs and understanding therapeutic mechanisms. In: An NIH White Paper by the QSP Workshop Group-October, vol. 2011 (2011)
31. Tian, F., Deguchi, K., Yamashita, T., Ohta, Y., Morimoto, N., Shang, J., Zhang, X., Liu, N., Ikeda, Y., Matsuura, T., et al.: In vivo imaging of autophagy in a mouse stroke model. Autophagy **6**(8), 1107–1114 (2010)
32. Tyson, J.J., Chen, K., Novak, B.: Network dynamics and cell physiology. Nat. Rev. Mol. Cell Biol. **2**(12), 908–916 (2001)
33. Tyson, J.J., Chen, K.C., Novak, B.: Sniffers, buzzers, toggles and blinkers: dynamics of regulatory and signaling pathways in the cell. Curr. Opin. Cell Biol. **15**(2), 221–231 (2003)
34. Tyson, J.J., Baumann, W.T., Chen, C., Verdugo, A., Tavassoly, I., Wang, Y., Weiner, L.M., Clarke, R.: Dynamic modelling of oestrogen signalling and cell fate in breast cancer cells. Nat. Rev. Cancer **11**(7), 523–532 (2011)
35. van der Greef, J., McBurney, R.N.: Rescuing drug discovery: in vivo systems pathology and systems pharmacology. Nat. Rev. Drug Discov. **4**(12), 961–967 (2005)
36. Volinia, S., Calin, G.A., Liu, C.G., Ambs, S., Cimmino, A., Petrocca, F., Visone, R., Iorio, M., Roldo, C., Ferracin, M., et al.: A microrna expression signature of human solid tumors defines cancer gene targets. Proc. Natl. Acad. Sci. USA **103**(7), 2257–2261 (2006)
37. Weissleder, R.: Scaling down imaging: molecular mapping of cancer in mice. Nat. Rev. Cancer **2**(1), 11–18 (2002)
38. Weissleder, R.: Molecular imaging in cancer. Science **312**(5777), 1168–1171 (2006)
39. Wu, S., Wang, X., Chen, J., Chen, Y.: Autophagy of cancer stem cells is involved with chemoresistance of colon cancer cells. Biochem. Biophys. Res. Commun. **434**, 898–903 (2013)

40. Xu, Y., Yuan, J., Lipinski, M.M.: Live imaging and single-cell analysis reveal differential dynamics of autophagy and apoptosis. Autophagy **9**(9), 1418–1430 (2013)
41. Zhao, S., Iyengar, R.: Systems pharmacology: network analysis to identify multiscale mechanisms of drug action. Annu. Rev. Pharmacol. Toxicol. **52**, 505 (2012)
42. Zhu, H., Wu, H., Liu, X., Li, B., Chen, Y., Ren, X., Liu, C.G., Yang, J.M.: Regulation of autophagy by a beclin 1-targeted microrna, mir-30a, in cancer cells. Autophagy **5**(6), 816–823 (2009)

# Chapter 5
# Source Code for Dynamic Model of Interplay Between Autophagy and Apoptosis

For solving the ODEs used for modeling autophagy and apoptosis pathway we used MatlabR2012. The Matlab source code is provided in this chapter:

The first file contains ODEs and algebraic equations:

```
%Autophagy Model

function dy = Autophagy_fun(t,y,ka,kda,C,kc,ksb,krb,kra,Gj,Gd,...
    Gb,Gt,Gg,Gh,Gl,Gc,sigmaj,sigmad,sigmab,sigmat,sigmag,...
    sigmah,sigmal,kcasp,kin,kout,wbec0,wjnk0,wdapk0,wbcl0,...
    watg0,wmtor0,wbh0,wjnk_s,wdapk_s,wbec_d,wbcl_j,...
    watg_m,wmtor_ca,wbh_ca,wbh_s,BCL2T,IP3RT,BCLmit,CaT)
BECNT = y(1);Atphg  = y(2);S = y(3);JNK  = y(4); DAPK  = y(5);...
    BCLP = y(6); mTOR  = y(7); Atg13  = y(8);
Ca  = y(9);BH3 = y(10); CASP = y(11); BECNP = y(12);
%Soft heaviside function
H=inline('1/(1+exp(-s*W))','W','s');
ustep = @(x) x>=0;%Step Function
astep = @(x) x>0;%Step Function

%Algebric Equations
BCL2U=BCL2T-BCLP;
BECNU=BECNT-BECNP;
LIGtotal=IP3RT+BECNU;
LIGfree= max(0,(LIGtotal-BCL2U));
BECNF=BECNU*LIGfree/LIGtotal;
IP3RF=IP3RT*LIGfree/LIGtotal;

W_D=wdapk0+wdapk_s*S;
W_J=wjnk0+wjnk_s*S ;
W_B=wbcl0+wbcl_j*JNK ;
W_T=wmtor0+wmtor_ca*Ca ;
W_H=wbh0+wbh_ca*Ca+wbh_s*S;
W_G=watg0+watg_m*mTOR ;
W_L=wbec0+wbec_d*DAPK ;
```

 
I. Tavassoly, *Dynamics of Cell Fate Decision Mediated by the Interplay of Autophagy and Apoptosis in Cancer Cells*, Springer Theses, DOI 10.1007/978-3-319-14962-2_5

```
%ODEs
dy(1)=-kcasp*CASP*BECNT;
dy(2)=ka*(BECNF+BECNP)*Atg13-kda*Atphg;
dy(3)=ksb+C*kc*ustep(t)-(krb+kra*Atphg)*S;
dy(4)=Gj*(H(W_J,sigmaj)-JNK);
dy(5)=Gd*(H(W_D,sigmad)-DAPK);
dy(6)=Gb*(BCL2T*H(W_B,sigmab)-BCLP);
dy(7)=Gt*(H(W_T,sigmat)-mTOR);
dy(8)=Gg*(H(W_G,sigmag)-Atg13);
dy(9)=kout*IP3RF*(CaT-Ca)-kin*Ca;
dy(10)=Gh*(H(W_H,sigmah)-BH3);
dy(11)=Gc*(astep(BH3-BCLmit)-CASP);
dy(12)=Gl*(BECNU*H(W_L,sigmal)-BECNP);%

dy=dy';
```

The second file contains parameters and initial conditions:

```
  %Parameters
ka=2;C=0;kda=0.3;
kc=0.25;ksb=0.1;krb=1;kra=10*kda;
Gj=1;Gd=1;Gb=0.5;Gt=1;Gg=1;Gh=0.2;Gc=1;Gl=10;
sigmaj=5;sigmad=10;sigmab=2;sigmat=10;sigmag=4;
sigmah=1;sigmal=1;
kcasp=2.5;kin=1;kout=0.5;
wbec0=-0.3;wjnk0=-0.5;wdapk0=-0.3;wbcl0=-0.8;
watg0=0;wmtor0=2;wbh0=-5;
wjnk_s=0.3;wdapk_s=0.1;
wbec_d=0.2;wbcl_j=1;
watg_m=-1;wmtor_ca=-10;
wbh_ca=11;wbh_s=0.1;
BCL2T=3;IP3RT=1;
CaT=2;
live=0;
LC3II=0;

%Random selection of BCL2mit from a normal distribution
bcl2m_vec=0.1+0.03.*randn(100,1);
 for s=1:length(bcl2m_vec)
BCLmit=bcl2m_vec(s);
options = odeset('RelTol',1e-12,'AbsTol',1e-12);
tspan=0:1:24;   %For BCL2 Overexpression case: tspan=-10:1:24
BECNT=3;Atphg=0.774;S=0.030;JNK=0.079;DAPK=0.046;BCLP=0.573;
mTOR=0.592;Atg13=0.085;Ca=0.196;BH3=0.055;CASP=0;BECNP=0.898;
y0 = [BECNT Atphg S JNK DAPK BCLP mTOR Atg13 Ca BH3 CASP BECNP];
[t,y] = ode15s(@Autophagy_fun,tspan,y0,options,ka,kda,C,kc,ksb,krb,...
    kra,Gj,Gd,Gb,Gt,Gg,Gh,Gl,Gc,sigmaj,sigmad,sigmab,sigmat,sigmag,...
    sigmah,sigmal,kcasp,kin,kout,wbec0,wjnk0,wdapk0,wbcl0,watg0,...
    wmtor0,wbh0,wjnk_s,wdapk_s,wbec_d,wbcl_j,watg_m,wmtor_ca,...
    wbh_ca,wbh_s,BCL2T,IP3RT,BCLmit,CaT);
```

```
 LC3II=y(:,2)/0.774+LC3II;
 %0.774 is basal level of autophagy for Cisplatin=0

 if y(13,11)==0 %Caspase activation at time=12
  live=live+1;
 end
end
 apoptosis=100-live;
 B=LC3II(:,1)/100;%Average of LC3II level for 100 Cells/
 Simulations
```

MIX
Papier aus verantwortungsvollen Quellen
Paper from responsible sources
FSC® C105338

If you have any concerns about our products,
you can contact us on
**ProductSafety@springernature.com**

In case Publisher is established outside the EU,
the EU authorized representative is:
**Springer Nature Customer Service Center GmbH**
**Europaplatz 3, 69115 Heidelberg, Germany**

Printed by Libri Plureos GmbH
in Hamburg, Germany